ELEMENTARY MECHANICS OF SOLIDS AND FLUIDS

MAXWELL PRESS

ELEMENTARY MECHANICS OF SOLIDS AND FLUIDS

Arthur Laidlaw Selby M.A.

Fellow of Merton College

MAXWELL PRESS

Chennai Trichy New Delhi

MAXWELL PRESS

This edition has been published in india by arrangement with Carson Books, UK

ISBN 978-93-5528-189-0

Maxwell Press
No. 44, Nallathambi Street,
Triplicane, Chennai 600 005

Printed and bound in India

MJP 1276

Publisher : C. Janarthanan

Publisher's Note

The legacy of a country is in its varied cultural heritage, historical literature, developments in the field of economy and science. The top nations in the world are competing in the field of science, economy and literature. This vast legacy has to be conserved and documented so that it can be bestowed to the future generation. The knowledge of this legacy is slowly getting perished in the present generation due to lack of documentation.

Keeping this in mind, the concern with retrospective acquiring of rare books has been accented recently by the burgeoning reprint industry. Maxwell Press is gratified to retrieve the rare collections with a view to bring back those books that were landmarks in their time.

In this effort, a series of rare books would be republished under the banner, "Maxwell Press". The books in the reprint series have been carefully selected for their contemporary usefulness as well as their historical importance within the intellectual. We reconstruct the book with slight enhancements made for better presentation, without affecting the contents of the original edition.

Most of the works selected for republishing covers a huge range of subjects, from history to anthropology. We believe this reprint edition will be a service to the numerous researchers and practitioners active in this fascinating field. We allow readers to experience the wonder of peering into a scholarly work of the highest order and seminal significance.

MAXWELL PRESS

PREFACE

I HOPE that this book may prove useful to students, who without possessing much knowledge of Mathematics desire to read Mechanics as an introduction to Physics. Those who are acquainted with the elements of Algebra and Geometry will, I think, meet with no serious mathematical difficulty; they may, however, find it convenient to read first the geometrical theorems at the end of Chapter I.

Some propositions on the geometry of the ellipse are given in Chapter VI, in order to render the discussion of the law of gravitation more complete.

My thanks are due to Mr. J. Walker, M.A., of Christ Church, for his kindness in reading the proof-sheets, and to Professor J. V. Jones, M.A., for many valuable suggestions.

A. L. S.

CONTENTS

CHAPTER I.

KINEMATICS.

CHAPTER II.

THE LAWS OF MOTION.

CHAPTER III.

WORK AND ENERGY.

CHAPTER IV.

MOTIONS OF AN EXTENDED BODY.

CHAPTER V.

MACHINES. PROBLEMS IN STATICS.

CHAPTER VI.

GRAVITATION.

CHAPTER VII.

ELASTICITY.

CHAPTER VIII.

HYDROSTATICS.

CHAPTER IX.

CAPILLARITY.

CHAPTER X.

UNITS AND THEIR DIMENSIONS.

CHAPTER IX.

CAPILLARITY.

ELEMENTARY MECHANICS.

CHAPTER I.

KINEMATICS.

§ 1. The Science of Mechanics investigates the conditions which govern the motion of bodies.

It comprises two parts, Kinematics and Kinetics.

The object of Kinematics is to describe and classify the changes in the motions of moving bodies. It is the province of Kinetics to assign the causes of these changes.

A body may remain at rest, or in equilibrium, either from the absence of causes which tend to move it, or because such causes balance one another.

The part of Mechanics which treats of the conditions under which bodies remain at rest, is a branch of Kinetics and is called Statics.

As we cannot hope to explain the phenomena of motion before we know how to classify them, we must first occupy ourselves with Kinematics. We shall therefore discuss in this chapter the simple phenomena of motion without reference to their causes.

The form and size of moving bodies will be supposed to remain unchanged during motion.

§ 2. Measurement of Time and Length.

Times and lengths (or distances) are expressed as multiples of convenient units.

The unit of time is generally the second. The unit of length is the centimetre or the foot. A time 5 means 5 seconds; a length 10 is 10 centimetres or 10 feet according as the centimetre or foot is the unit.

The centimetre (0·3937079 inch) is the International Scientific Unit; the foot is only used in Britain.

Other units might be chosen, as the inch and the minute. Expressed in terms of these units, 5 seconds and 10 feet would be denoted by $\frac{1}{12}$ and 120 respectively.

Derived Units. Square Measure. Cubic Measure.

The most convenient unit of area is the area of the square on the unit of length, for the area of an oblong of length a and breadth b is denoted by ab, when measured in terms of this unit.

Similarly the unit of volume is the volume of a cube, each edge of which is of unit length.

As the units of area and volume depend on the unit of length, they are called derived units.

The centimetre, square centimetre, and cubic centimetre, are denoted by cm., sq.cm., and c.cm.

§ 3. **Displacement.**

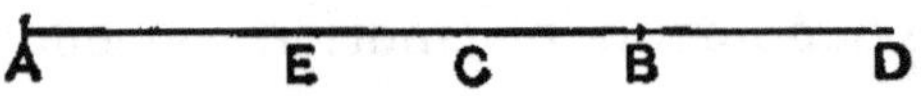

A E C B D

Fig. 1.

When a body is displaced so that a point of it is moved from A to B, the change of position of the point is described by saying that it has a displacement AB.

The displacement is only completely described when the magnitude and direction of AB are given. The magnitude of the displacement is measured by the number of units of length in AB. If the direction of the displacement is not also given, B may be anywhere on a sphere with centre A and radius AB.

Quantities, such as displacement and velocity, which are

only completely defined when their magnitude and direction are both known, are called Vectors.

A vector whose magnitude is *AB*, and whose direction is that of the line from *A* to *B*, can be represented by this line, and is called the vector *AB*.

Equal vectors are those which are represented by equal and parallel straight lines drawn towards the same parts.

If a point has successive displacements in the same direction the total displacement is the sum of these displacements.

If the displacements are all along the same straight line, as *AB*, *BC*, *CD*, *DE*, some in one direction some in the other, displacements in opposite directions must be given opposite signs, and the algebraic sum of all is the total displacement in the positive direction.

For example, displacements of 3 cm. and 4 cm. to the right, combined with a displacement of 8 cm. to the left, make a displacement of —1 to the right, or 1 to the left.

A body is said to have a displacement of translation, when every point of it has the same displacement.

§ 4. Composition of displacements.

Let successive displacements *AB*, *BC* be given to a body; then since a point originally at *A* is finally at *C*, the total displacement is *AC*.

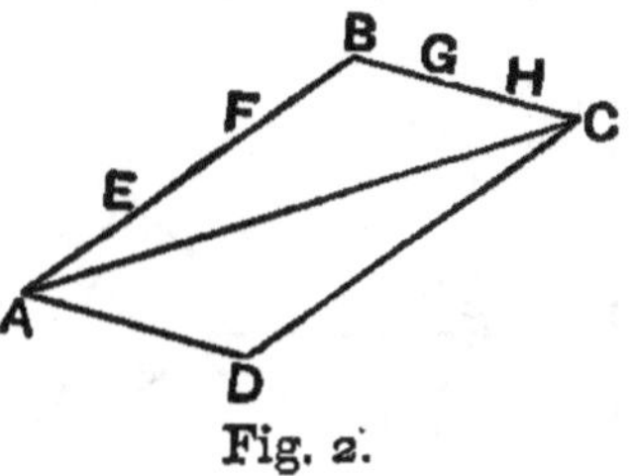

Fig. 2.

The final position of the body displaced depends only on the magnitude and direction of the displacements *AB*, *BC*, and not on the order in which they take place. Indeed we may break up each displacement into parts, as *AE*, *EF*, *FB*, and *BG*, *GH*, *HC*, and communicate them to the body in any order we please; the result is still to produce the same displacement *AC*.

AC is called the resultant displacement, and *AB*, *BC* are called its components. The rule for finding the resultant displacement when the components are given is called the Parallelogram Law, and is as follows.

Let a parallelogram be constructed whose adjacent sides, AB, AD, represent the magnitudes and directions of the component displacements.

The diagonal AC, which is drawn from the angle in which these sides meet, represents the magnitude and direction of the resultant displacement.

Conversely, a displacement AC may be replaced by displacements AB, BC or AB, AD, where AB is any straight line through A. AC is then said to be resolved into its components AB, BC or AB, AD.

The most useful case of resolution is that in which ADC is a right angle.

In this case if $AC = R$ and $\angle CAD = a$, the component displacements x, y, along AD, DC respectively, are $R \cos a$, $R \sin a$; and

$$x^2 + y^2 = R^2 ; \quad y = x \tan a. \tag{1}$$

Polygon of Displacements.

If a body has several successive displacements AB, BC, CD, DE, not necessarily all in the same plane, the resultant displacement is AE; and if the lines representing the displacements form a closed figure, the resultant displacement is zero.

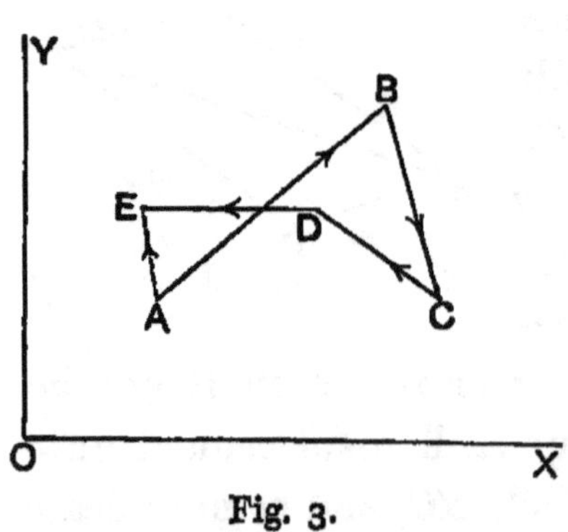

Fig. 3.

This proposition is called the Polygon of displacements.

The magnitude and direction of the resultant of several displacements, which are all in the same plane, is best found as follows.

Take two convenient lines OX, OY in the plane perpendicular to one another, and resolve each displacement into its components parallel to these.

Then if x be the sum of the components parallel to OX, and y the sum of the components parallel to OY, the magnitude of the resultant displacement is $\sqrt{x^2+y^2}$ and the angle a which it makes with OX is given by

$$\tan a = \frac{y}{x}.$$

Relative displacement.

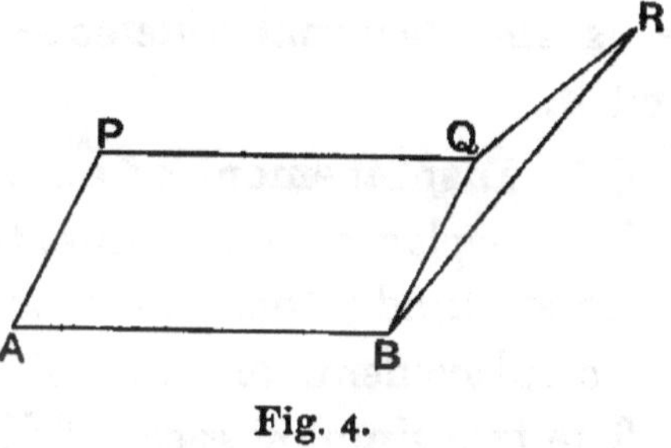

Fig. 4.

First, let two points A and B have equal displacements AP, BQ.

Then by Euclid I. 33, PQ is equal and parallel to AB, and since there is no change in the length or direction of the line joining the points, each is said to have no displacement relatively to the other.

Next, let the displacements be AP, BR.

Resolve BR into BQ and QR, of which BQ is equal to AP. QR is called the displacement of B relatively to A; it is compounded of QB, BR—that is, of the displacement of B, and the reversed displacement of A.

Also QR and PQ compound into PR.

Therefore the displacement of B relatively to A, and the original vector from A to B, compound into the vector from the final position of A to that of B.

The displacement of A relatively to B is equal in magnitude but opposite in direction to the displacement of B relatively to A.

If the displacements AP, BR are along the same straight line, the displacement of B relatively to A is the algebraic difference of the two displacements; it is also the algebraic sum of the displacement of B and the reversed displacement of A. If we call the resultant of two vectors their vectorial sum, we may appropriately use the term vectorial difference to denote the resultant obtained when one vector is reversed.

When the two displacements are in the same straight line, the algebraic difference is the vectorial difference.

Hence in every case the displacement of B relatively to A is the vectorial difference of the displacements of B and A.

§ 5. **Displacement of an extended body.**

The displacements of an extended body may be of a very complex kind; there may be no simple relation between the displacements of its points.

The two simplest cases of displacement are

(1) Displacement of translation; here all points of the body have the same displacement both in magnitude and direction, and the displacement of any point may be considered as that of the body.

(2) Displacement of rotation; here all points on one straight line (either in the body or fixed relatively to it)* remain fixed, and the displacement can be performed by rotation about this line as axis.

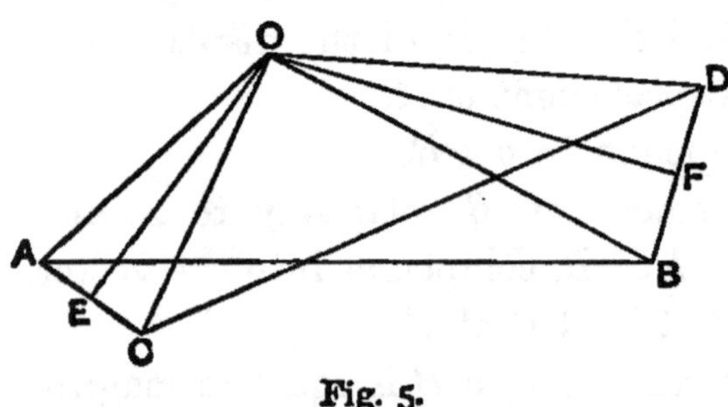

Fig. 5.

If the body is a plane figure which rotates in its plane,

* A point is said to be fixed relatively to a body when its distances from the several points of the body do not change. The straight line joining two such points is said to be fixed relatively to the body.

one point in the plane remains fixed, and the displacement may be described as rotation round this point.

Any displacement of a plane figure of invariable form in its own plane can be made either by a motion of translation or by a motion of rotation round an axis perpendicular to the plane.

Let A, B be the positions of two points of the figure before displacement; C, D their displaced positions: then the final position of any other point of the figure can be found.

Join AC and BD and bisect them at E and F.

Draw EO and FO at right angles to AC and BD respectively.

They will meet at O unless they are parallel, i.e. unless AC and BD are parallel.

But if AC and BD are parallel, the displacement can be made by a motion parallel to AC.

If AC and BD are not parallel, join OA, OB, OC, OD.

Then since $AE = EC$ and EO is perpendicular to AC, $OA = OC$.

Similarly $OB = OD$.

Also in the triangles AOB, COD, the two sides AO, OB are equal to the two sides CO, OD.

And $AB = CD$.

Therefore the angle AOB is equal to the angle COD.

And taking away the angle COB, the angles AOC, BOD are equal.

Therefore by rotation about O through an angle AOC, the point A is displaced to C, and B to D.

Since all lines in the plane are turned through the angle AOC, AOC is called the angular displacement.

NOTE.—It is supposed that in the displacement, the figure never quits the plane. A displacement such as that which a leaf of a book undergoes when it is turned over is excluded.

Two equal successive angular displacements of a body in opposite directions about parallel axes are equivalent to a displacement of translation.

Let the axes be perpendicular to the paper, meeting it in A and B.

If θ be the angular displacement, any line perpendicular to the axes is first turned through an angle θ and then through an angle $-\theta$. Hence it is parallel to its original position.

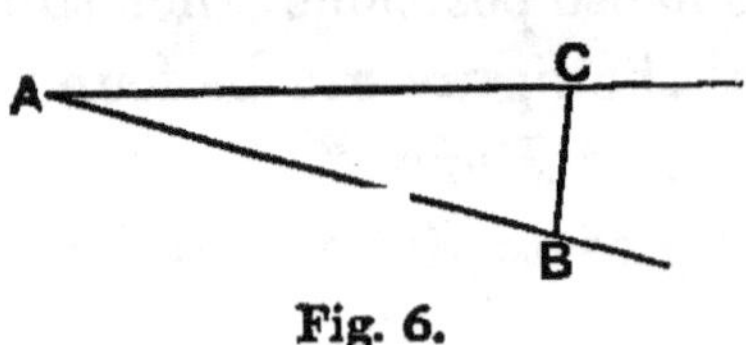

Fig. 6.

The displacement of translation is equal to BC the displacement of B. If $AB = l$, $BC = 2l \sin \frac{\theta}{2}$.

This result may also be stated thus:

An angular displacement θ about any axis can be replaced by an equal angular displacement about a parallel axis, at distance l, together with a displacement of translation $2l \sin \frac{\theta}{2}$ equally inclined to the initial and final positions of the shortest distance between the axes.

Examples.

1. Find the resultant of the following displacements, 3 N, 5 E, 6 S, and 9 W.

3 N and 6 S are equivalent to 3 S, and 5 E and 9 W are equivalent to 4 W.

Therefore if R be the resultant displacement, its direction lies between S and W and the magnitude of R is $\sqrt{3^2+4^2}$ or 5.

$\frac{3}{4} = \tan \alpha$ where α is the angle which R makes with the W.

An angle α whose tangent is x is often denoted by $\tan^{-1} x$.

Thus here $\alpha = \tan^{-1} \frac{3}{4}$.

The resultant can also be constructed on a diagram by drawing AB, BC, CD, DE to represent the magnitudes and directions of the given displacements. AE then represents the resultant.

2. Find the magnitude and direction of the resultant of the displacements 16 SW, $4\sqrt{2}$ N, and 9 SE.

The most convenient directions in which to resolve are SW, and SE.

$4\sqrt{2}$ N resolves into 4 NW and 4 NE, or into -4 SE and -4 SW.
Therefore the resultant displacement R is compounded of

$$16-4 \text{ or } 12 \text{ SW, and } 9-4 \text{ or } 5 \text{ SE.}$$

Therefore $R=\sqrt{12^2+5^2}=13$.

The direction of R lies between SW and SE, making an angle $\tan^{-1}\frac{5}{12}$ with the SW.

R may also be constructed on a diagram as in Ex. 1.

3. Find the resultant of the displacements

$$30 \text{ N}, 6\sqrt{2} \text{ SE, and } 1 \text{ W.}$$

4. A point has successive displacements 5, 10, 10 parallel to the sides AB, BC, CA of an equilateral triangle ABC taken in order; find the direction and magnitude of the resultant displacement.

Let AD be perpendicular to BC, and resolve along DA and DB.

5 along AB resolves into $\frac{5\sqrt{3}}{2}$ along AD and $\frac{5}{2}$ along DB.

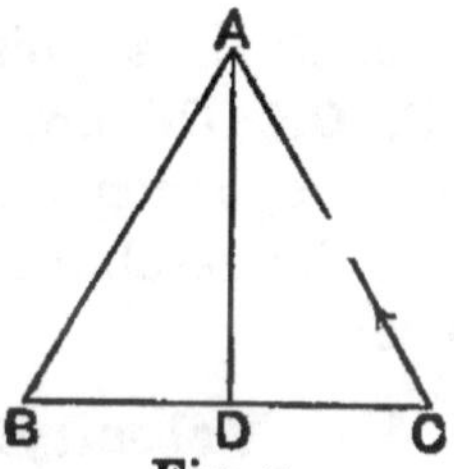

Fig. 7.

10 along CA resolves into $\frac{10\sqrt{3}}{2}$ along DA and 5 along CD.

Therefore the components are

$$\frac{10\sqrt{3}}{2}-\frac{5\sqrt{3}}{2}=\frac{5\sqrt{3}}{2}, \text{ along } DA.$$

$$\text{And } \frac{5}{2}+5-10=-\frac{5}{2}, \text{ along } DB$$

$$\text{or } \frac{5}{2}, \text{ along } DC.$$

The resultant is $\sqrt{\left(\frac{5}{2}\right)^2+\left(\frac{5\sqrt{3}}{2}\right)^2}$ or 5.

And it makes an angle $\tan^{-1}\sqrt{3}$ or 60° with DC.
Therefore the resultant displacement is 5 parallel to BA.

We can also find the resultant as follows; displacements 10, 10, 10 have a resultant zero; and therefore the given displacements have a resultant -5 parallel to *AB* or 5 parallel to *BA*.

5. *ABCD* is a square, and *O* is the intersection of its diagonals. Find the resultant of displacements $\sqrt{2}$, 2, 4, 4, $5\sqrt{2}$ respectively parallel to *OA, AB, BC, CD, DO*.

6. Find the resultant of displacements 1, 8, 3, 4, 5, 6 parallel to the sides of a regular hexagon taken in order.

7. The resultant of two equal displacements is equally inclined to each of them.

8. *ABC* is an isosceles triangle right-angled at *C*. Find the resultant of three equal displacements *a* respectively parallel (1) to *AB, BC, CA*; (2) to *AB, BC, AC*.

9. *D, E, F* are the middle points of the sides *BC, CA, AB* of any triangle *ABC*. Prove that the resultant of displacements equal and parallel to *AD, BE, CF* is zero.

10. From a point *O* within a triangle *ABC*, straight lines *OD, OE, OF* are drawn, which are perpendicular in direction and proportional in length to *BC, CA, AB* respectively. Show that the resultant of displacements represented by *OD, OE, OF* is zero.

§ 6. Velocity.

The displacement of a point from *A* to *B* can only take place by motion along a continuous path between *A* and *B*, and must occupy time.

The time occupied along the path will be longer or shorter according as the moving point travels more or less quickly; hence the rate at which a displacement proceeds is an important matter.

To take the simplest case, let the path be a straight line along which the moving point travels equal distances in equal times, however small.

The rate of displacement of the point is called its Velocity, and is measured by the distance travelled in unit time.

If s be the distance travelled in time t, $\frac{s}{t}$ is the distance travelled in unit time.

Therefore if v be the velocity,

$$v = \frac{s}{t}.$$

It has been noticed that s means s units of length; similarly v means v times a particular velocity which is called the unit.

Making $s = 1$, and $t = 1$, we have $v = 1$.

Therefore the unit of velocity is the velocity of a point which moves through unit distance in unit time.

Since this unit is determined when the units of length and time are known, it is a derived unit.

A velocity, like a displacement, is only fully represented when its direction as well as its magnitude is known. It is therefore a vector.

The formula $s = vt$ applies to motion with velocity of constant magnitude, whether the path is straight or curved, provided that s is measured along the path.

Examples.

1. Express a velocity of 30 miles an hour in feet per second.

$$30 \text{ miles} = 30 \times 5280 \text{ feet}.$$
$$1 \text{ hour} = 3600 \text{ seconds}.$$

Therefore $v = \frac{30 \times 5280}{3600} = 44$ feet per second.

2. A train travels $22\frac{1}{2}$ miles an hour. How far does it go in 50 seconds?

3. A cyclist rides at a rate of 25 feet per second. How long does he take to cover a mile?

When a body moves over unequal distances in equal times, the magnitude of its velocity varies.

The conception of quantities which vary with the time is very important in Physics. It is first necessary to explain the term 'instant of time.'

An instant of time is analogous to a point of space; as a point has no magnitude, so an instant has no duration; e. g. if a circular disc be divided along a diameter into semicircles coloured black and white respectively, and a pointer of the form of a sector of the disc revolve continuously round its centre, there is a definite instant at which the forward edge of the pointer passes from the black to the white part of the disc, and a subsequent instant at which the hinder edge does the same. These instants are separated by an interval of time to which we must attach the idea of magnitude. No such idea is attached to an instant.

§ 7. Diagrams.

Let there be some physical quantity, as the height of

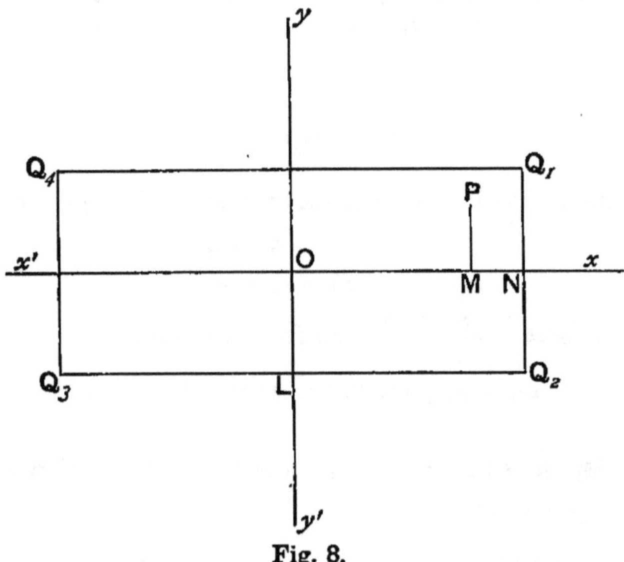

Fig. 8.

a barometer or the temperature of a room, which at any

instant has a definite magnitude, but may vary from time to time; the law of variation of such a quantity may be indicated on a diagram.

Draw two perpendicular lines $x'Ox$, $y'Oy$; and from any point P draw PM parallel to Oy. OM is called the abscissa, PM the ordinate of P. If a point lies above $x'Ox$, its ordinate is considered positive; if below $x'Ox$, the ordinate is negative.

The abscissa of a point is positive or negative, according as the point is to the right or to the left of $y'Oy$.

The position of any point is known when the magnitude and sign of the ordinate and abscissa are given. For let the abscissa and ordinate be a, $-b$, respectively. Along Ox take ON equal to a, and draw NQ_2 downwards equal to b.

Then Q_2 is the point required.

If $Q_1\ Q_2\ Q_3\ Q_4$ be a rectangle with its sides parallel to Ox, Oy, and its diagonals pass through O, $Q_1N = NQ_2$ and $Q_2L = LQ_3$.

A point of which the ordinate and abscissa are x, y, is often denoted by (x, y).

The points Q_1, Q_3, Q_4, are respectively

$$(a, b), (-a, -b), (-a, b).$$

Let there be two quantities A and B, which are mutually dependent, so that A varies when B does, and A can be found when B is known.

Take any number of points $a, b, c, \ldots$ along Ox; Oa, Ob, Oc, ... being possible values of A. Through $a, b, c \ldots$ draw ap, bq, cr ... parallel to Oy, and equal respectively to the values which B has when A has the values Oa, Ob, Oc, ...

Then there is a curve passing through $p, q, r, \ldots$ such that the abscissa of any point of it gives the value of A, and the ordinate the corresponding value of B; and when this curve is drawn the relation between A and B is completely expressed.

A familiar instance of such a curve is given by a

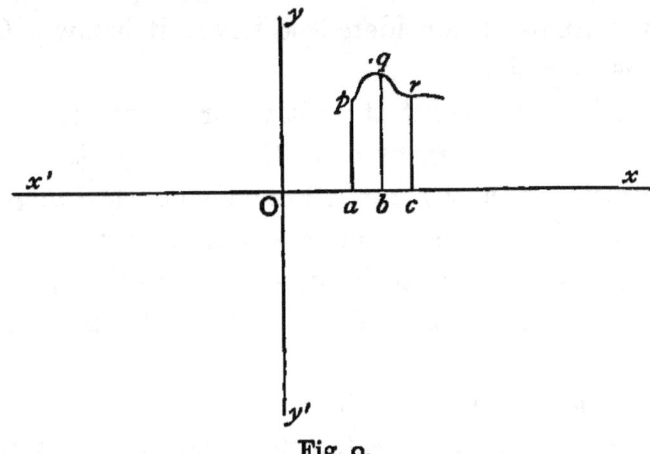

Fig. 9.

barometer chart; the abscissæ measured in divisions on the paper represent the number of hours or days that have elapsed since a given instant, and the ordinates represent on a convenient scale the corresponding heights of the barometer.

If A is negative, the curve will lie on the left of yOy', above Ox' when B is positive, below Ox' where B is negative. If A is positive and B negative, the curve lies within the angle xOy'.

§ 8. Application to velocity. Curve of Positions.

Let a diagram be drawn on which the abscissæ are the times reckoned from a certain instant which is called the Epoch, and the corresponding ordinates are the distances traversed by a moving point in these times.

Thus in fig. 10 let PM, QN denote the distances traversed in the times OM, ON, by a point. The curve RPQ representing the motion of this point is called the curve of Positions.

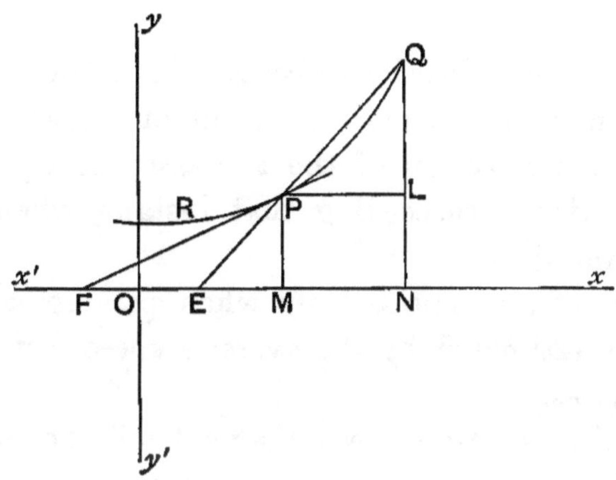

Fig. 10.

Draw PL parallel to Ox, and produce QP to meet Ox in E. Draw PF to touch the curve at P.

Then QL is the distance traversed in time MN, and $\frac{QL}{MN}$ is the average velocity during this time,

$$\frac{QL}{NM} = \frac{QL}{LP} = \frac{PM}{ME} = \frac{PM}{MF - EF}.$$

Now let Q move along the curve towards P, and ultimately coincide with it; then QP tends more and more to coincide with the tangent at P.

But as Q approaches P, $\frac{EF}{MF}$ diminishes indefinitely, and vanishes when Q coincides with P.

Thus when the time MN is diminished indefinitely, the average velocity approaches and ultimately has the value $\frac{PM}{MF}$, for it differs from this by less than any assigned quantity.

$\frac{PM}{MF}$ is therefore said to be the velocity of the moving point at the instant M, or at a time OM reckoned from the Epoch.

Measurement of Variable Velocity. Thus the magnitude and direction of a variable velocity at any instant, are the magnitude and direction of the average velocity during a time t immediately succeeding that instant, when t is indefinitely diminished.

The speed of an express train when passing a point P could not be estimated by the average speed between two stopping-places.

If the train were timed over the next mile beyond P, the speed at P would sometimes be determined with fair accuracy, sometimes not. But if the time taken to traverse a yard or foot from P could be accurately found, this would give the velocity at P with great accuracy, for in so short a time the velocity would not vary much.

Our system of measuring variable velocity is devised on this plan, but the time of the measured motion is made exceedingly small, and the accuracy of measurement can be increased as much as we please by diminishing this time.

The curve of positions has been drawn as a continuous curve with no sharp points; this covers all cases that we need consider.

The tangent to the curve of positions is only parallel to Ox for instants when the point is at rest.

When the velocity is uniform and equal to v, the curve of positions is a straight line inclined to Ox, at an angle $\tan^{-1} v$.

§ 9. Curve of Velocity.

Let us now take another diagram on which the abscissæ represent times as before, but the ordinates are the corresponding velocities of a moving point. Any given motion of a point will be represented by a curve which we shall call the curve of velocity.

If the velocity is uniform and equal to v, the curve is a straight line PQ parallel to Ox at a distance v from it.

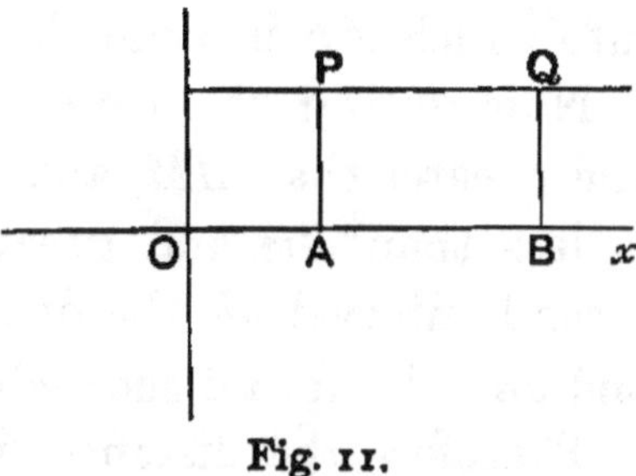

Fig. 11.

Let A and B denote instants of time separated by an interval t. Then, since the velocity is uniform, the distance travelled in time t is vt or $AP.AB$, that is, the area $PABQ$ contained between the axis of x, the curve of velocity, and the ordinates at A and B.

The same statement holds when the velocity varies in any manner.

For let MN be the curve of velocity, and let AB (fig. 12) be divided into any number of parts such as Aa, ab, bc, cB, each equal to d.

Through A, a, b, c, B draw ordinates meeting the curve in M, m, n, p, N. Complete the rectangles Mm, mn, np, pN, by parallels to Ox produced to meet BN in f, g, h, k.

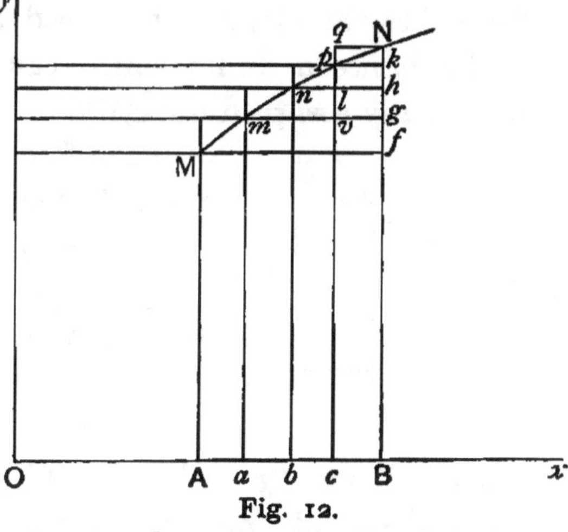

Fig. 12.

The rectangles vf, Mm are equal, for they are on equal bases and between the same parallels.

For the same reason the rectangles mn, lg are equal, and so are pn, kl.

Therefore the rectangle fq whose base is d and altitude

$BN - AM$ is equal to the sum of the rectangles Mm, mn, np, and pN, whose diagonals are the successive chords between M and N.

This statement holds, whatever be the number of parts into which AB is divided.

Now during the time Aa, the velocity is less than am and greater than AM, and therefore the distance traversed is less than Am and greater than aM. Similarly in the second interval ab, the distance traversed lies between bm and an. So for all succeeding intervals.

Therefore the distance traversed in the whole time AB is less than the sum of the areas Am, an, bp, cN, and greater than the sum of aM, bm, cn, Bp.

And the area bounded by the curve MN, the axis Ox, and the ordinates AM, BN lies within the same limits.

Therefore the distance traversed and the area $AMNBA$ lie within limits which differ by fq or by $d\,(BN - AM)$.

Now by increasing the number of equal intervals between A and B, d can be diminished indefinitely, and $d\,(BN - AM)$ can be made less than any assigned quantity.

Therefore the distance traversed and the area described both lie between limits which can be made to differ by less than any assigned quantity.

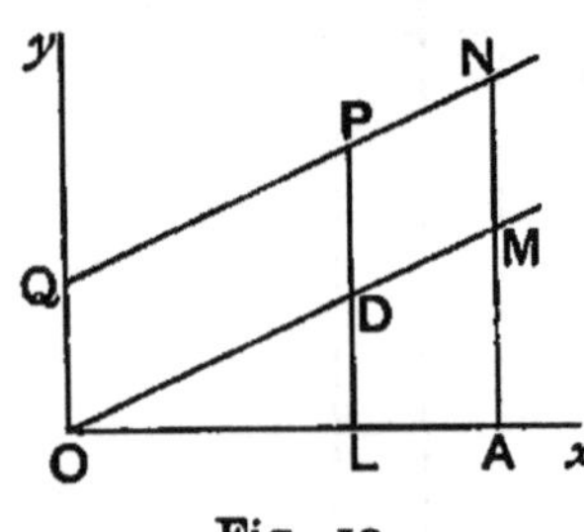

Fig. 13.

Therefore the area $AMNBA$ represents the distance traversed.

Uniformly increasing Velocity.

Let the velocity increase uniformly from v_0 to v in a time t.

Make OQ and OA equal to v_0 and t respectively, and through A draw AMN parallel to Oy, and make $AN = v$.

Join QN, and from any point P on QN draw PDL parallel to Oy. Draw OM parallel to QN.

Let $OL = x$.

Then $PL = PD + DL = OQ + \frac{x}{t} AM = v_0 + \frac{x}{t}(v - v_0)$.

Now $\frac{x}{t}(v - v_0)$ is the velocity acquired in a time x and v_0 is the initial velocity.

Therefore PL represents the velocity after a time x, and P is any point on QN. Therefore QN is the velocity-curve.

The area $OQNA = \frac{1}{2} OA (OQ + NA) = \frac{1}{2}(v_0 + v) t$.

Therefore $s = \frac{1}{2}(v_0 + v) t$.

Whence s is the distance travelled in time t.

If the body starts from rest $v_0 = 0$,

$$\text{and } s = \tfrac{1}{2} vt,$$

where v is the final velocity.

§ 10. Composition and Resolution of Velocities.

Let PQ represent the velocity of a moving point.

Draw PE, QE parallel to two lines Ox, Oy, and meeting at E.

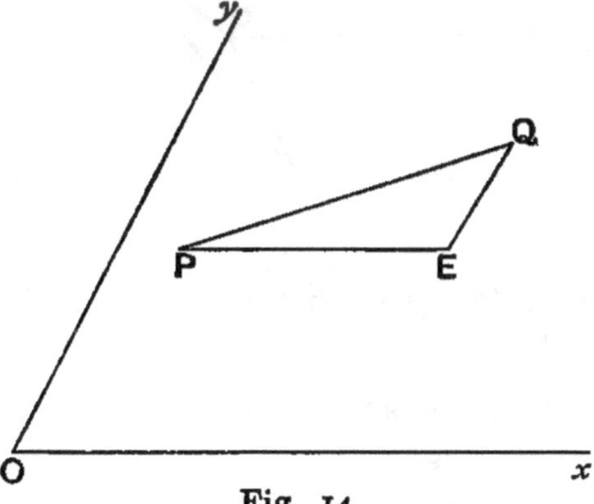

Fig. 14.

Then PE, EQ are the components parallel to Ox, Oy of the displacement in unit time. They are therefore called the components of the velocity along Ox and Oy; and if PE, EQ are known, the actual or resultant velocity can be obtained by compounding them according to the parallelogram law (§ 4).

If xOy is a right angle, R the magnitude of the velocity

PQ, a the angle between PQ and Ox, X and Y the components along Ox and Oy,

$$X = R \cos a, \quad Y = R \sin a,$$

$$R^2 = X^2 + Y^2, \quad \tan a = \frac{Y}{X}.$$

Relative Velocity.

If two points A and B are moving with the same velocity, the line AB remains unchanged in magnitude and direction; B is then said to have no velocity relatively to A.

Next, let the velocities be different, AP and BR (fig. 15) representing the displacements in unit time.

Then QR is the displacement of B relatively to A in unit time. This is called the velocity of B relatively to A, and is obtained by compounding the velocity of B with that of A reversed.

The velocity of B is obtained by compounding the velocity of B relatively to A with the velocity of A.

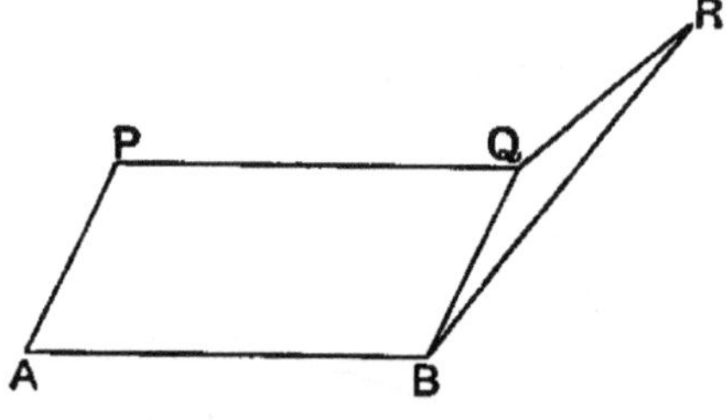

Fig. 15.

Similarly, if there are three moving points A, B, C, the velocity of C is compounded of the velocity of C relatively to B, the velocity of B relatively to A, and the absolute velocity of A.

Thus velocities are compounded according to the same law as displacements.

The idea of relative motion is of great importance in Mechanics, for we do not know any fixed point in space, or the absolute velocity of any point. The true velocity of a falling stone when it begins its course is not merely that which the thrower imparts to it; with this there are compounded the velocity of the point of projection round the earth's axis and the velocity of the earth's centre as it travels through space.

The motions of any number of points relatively to one another are not affected if their actual velocities are compounded with an additional velocity which is the same for all points.

A body is said to have a velocity of translation when the velocities of all points of it are the same; the velocity of any point of it is then the velocity of the body.

Examples.

1. A steamer goes $7\sqrt{2}$ miles an hour N. in a S.E. wind blowing 6 miles an hour. From what direction and with what velocity does the wind appear to blow to a passenger on the steamer.

Impress a velocity of $7\sqrt{2}$ miles an hour S. on both the wind and the steamer. The relative motion is unaltered, the steamer is brought to rest, and we have to compound velocities 6 N.W. and $7\sqrt{2}$ S. supposing the mile and hour are units.

6 N.W. $= 3\sqrt{2}$ N. and $3\sqrt{2}$ W.

Therefore the relative velocity is compounded of $(7-3)\sqrt{2}$ S. and $3\sqrt{2}$ W.

Therefore the velocity required is $5\sqrt{2}$, directed along a line OM in the S.W. quadrant which makes an angle $\tan^{-1}\frac{4}{3}$ with OW.

The wind then appears to blow almost from the N.E.

2. *A* who walks at a rate of 3 miles an hour starts 13 minutes before *B*, who walks on the same road at a rate of 4 miles an hour. When will *B* overtake *A*?

Impress a speed of -3 miles per hour on both *A* and *B*. Their relative motion is not affected.

Also in 13 minutes *A* walked $\frac{13}{20}$ mile.

Hence the question becomes—How long will *B* take to walk $\frac{13}{20}$ mile at the rate of one mile an hour?

3. The extremities *P* and *Q* of the hour and minute hands of a clock are distant 2 and 4 inches respectively from the centre of

the face. Find the velocity of Q relatively to P at 12, and at 3 o'clock, in foot-second units.

Let x and y be the velocities of P and Q.

The point Q traverses a distance $\frac{2\pi}{3}$ in 3600 seconds.

Therefore $y = \frac{22 \times 2}{7 \times 3 \times 3600} = \cdot000146.$

And $\quad x = \frac{22 \times 2}{7 \times 6 \times 12 \times 3600} = \frac{1}{24} y.$

The velocity of Q relatively to P is found by compounding y and $-x$.

Therefore the required velocity at 12 o'clock is $\frac{23}{24} y$.

At 3 o'clock x is directed vertically downwards.

Hence $-x$ is vertically upward.

And if R be the required velocity at 3 o'clock, a the angle it makes with the horizon,

$$R = \cdot000146 \sqrt{1 + \frac{1}{24^2}} = \frac{\cdot000146}{24} \sqrt{577},$$

$$\tan a = \frac{1}{24}.$$

4. In the last example, find the velocity of Q relatively to P at 1.30 and at 10.30.

5. A boat is rowed at the rate of 4 miles an hour across a river, a quarter of a mile broad, being steered in a direction at right angles to the current. If the velocity of the stream is 3 miles an hour, find the resultant velocity of the boat, and the distance below the starting point at which it reaches the opposite side.

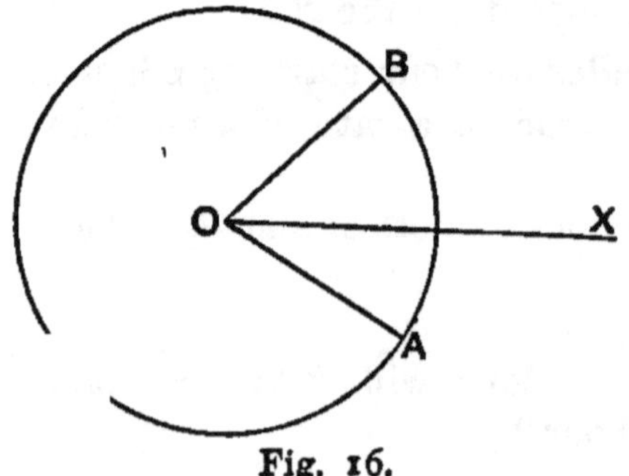

Fig. 16.

Angular Motion. Definition of a Particle.

Let a body, free to turn about an axis through O perpendicular to the plane of the paper, be displaced so that a straight line in it which initially coincides with OA finally coincides with OB.

Let $\angle AOB = d$, $AO = r$, arc $AB = s$.

Then $d = \dfrac{s}{r}$.

If the body turns through equal angles in equal times, it is said to have uniform angular velocity.

If t is the time occupied in the displacement d, $\dfrac{d}{t}$ is the angular displacement in unit time, and is called the angular velocity.

Denoting this by ω we have $d = \omega t$.

And $r\omega = \dfrac{rd}{t} = \dfrac{s}{t}$.

Therefore if v is the velocity of a point A, distant r from the axis,

$$v = r\omega.$$

When the angular velocity is variable, the displacement d, by which it is measured, must be made very small.

If the angular velocity increases uniformly from ω_0 to ω in a time t, the velocity of A increases uniformly from $r\omega_0$ to $r\omega$, and (§ 9) the point A traverses a distance

$$\tfrac{1}{2}\,(r\omega_0 + r\omega)\,t.$$

Therefore the angular displacement d in a time t is given by

$$d = \tfrac{1}{2}(\omega_0 + \omega)\,t.$$

Let the point O have a velocity V along OX.

The velocity of A is then compounded of V along OX and of $r\omega$ perpendicular to OA. The velocities of different points in the body are not the same. But if the body is made very small, r is very small, and $r\omega$ may be insensible compared with V; in this case all points of the body have the same velocity V.

A body of such small dimensions that the velocities of all its points are practically the same is called a Particle.

§ 11. Acceleration.

The rate at which the velocity of a point changes is called its Acceleration. If v be the velocity acquired in a time t, $\frac{v}{t}$ is the velocity gained in unit time, supposing that the rate of gain is uniform.

Denoting acceleration by a, we have

$$a = \frac{v}{t},$$

$$\text{or } v = at.$$

a is a multiple of some unit. Since $a = 1$ when $v = 1$ and $t = 1$, the unit is the acceleration of a body which gains unit velocity in unit time.

This unit depends on the units of length and time, and is consequently a derived unit.

If the acceleration varies, the time t must be made indefinitely short.

The direction of the acceleration is that of the acquired velocity.

Acceleration is therefore a vector.

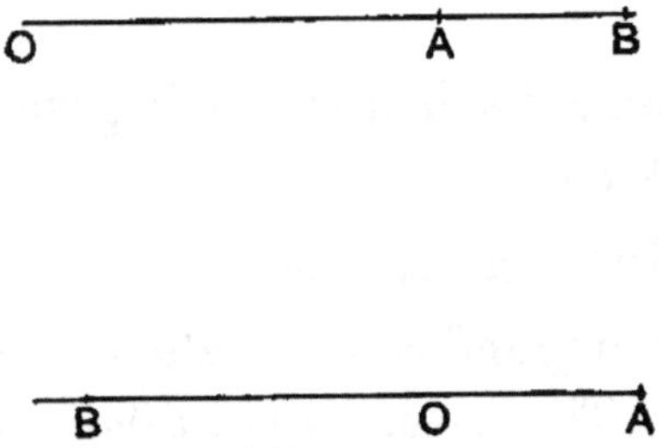

O A B

B O A

Fig. 17.

When the moving point possesses velocity at the beginning of the time t, two cases arise, according as the body moves along a straight line or not.

(1) Let v_0 and v be the initial and final velocities of the moving point represented by OA, OB.

Their difference is AB and the acceleration is $\frac{AB}{t}$ or $\frac{v - v_0}{t}$, provided that if this vary with the time, t is made very small. If $v < v_0$ the acceleration is negative.

(2) Let the initial and final velocities of the moving point P be in different directions as OA, OB (fig. 18).

Consider another point Q which moves with constant velocity OA.

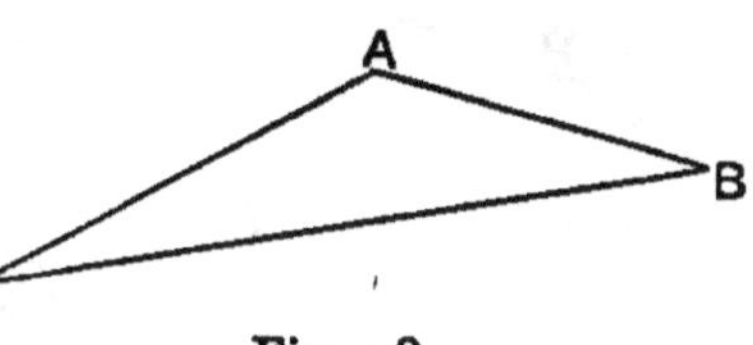

Fig. 18.

The velocity gained by P in the time t is its velocity relatively to Q at the end of the time.

This is AB (§ 10) and the acceleration is $\frac{AB}{t}$, its direction being along AB.

If either the magnitude or direction of the acceleration changes, t must be made indefinitely small.

A point moving on a curve necessarily has acceleration transverse to the direction of motion. For the path would be a straight line if the acceleration were only in the direction of motion.

Thus if AB be the curve, the velocities at A and B are directed along the curve and are not parallel. Therefore if the motion be from A to B a velocity towards the concave side of the curve must be compounded with the velocity at A in order to give a velocity along the tangent at B.

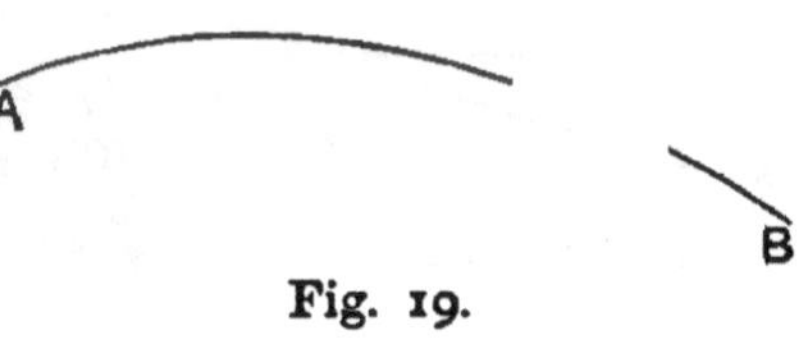

Fig. 19.

The Hodograph.

The position of a moving point at any instant is completely defined when the magnitude and direction of the vector drawn to it from a given fixed point O (fig. 20) are known.

If for successive instants the vectors indicating the corresponding positions of the point are drawn, the curve passing through the extremities of the vectors is the Path of the point.

Let another diagram (fig. 20 a) be taken on which the velocity of the moving point at any instant is represented in magnitude and direction by a vector from a fixed point o; and let vectors be drawn representing the velocity at successive instants.

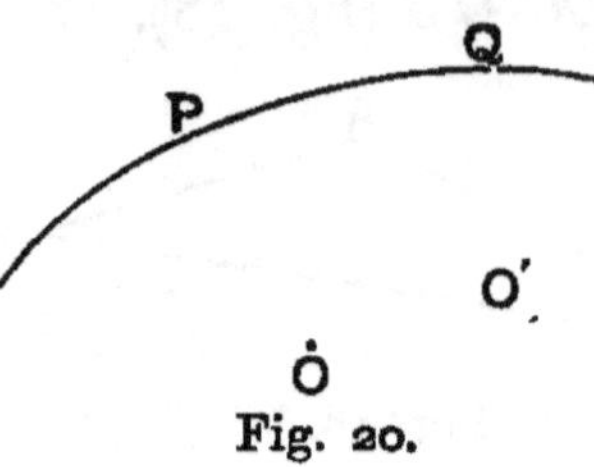

Fig. 20.

The curve passing through the extremities of these vectors is called the Hodograph of the point.

To each point (as P) on the path corresponds a point (as p) on the hodograph.

If t is the time occupied in passing from P to Q, $\frac{PQ}{t}$ is the average velocity between P and Q; and when Q moves up to P, the line PQ coincides with the tangent at P.

Hence the tangent to the path at P is parallel to op.

In the hodograph oq is compounded of op and pq.

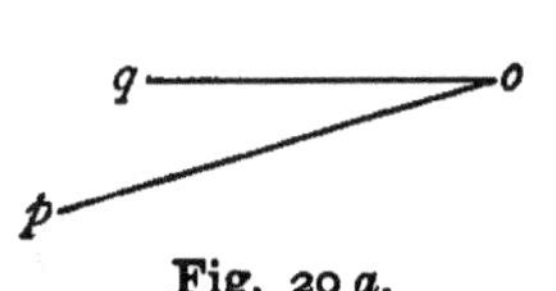

Fig. 20 *a*.

Therefore the velocity gained between the points P and Q of the path is pq, and the acceleration is $\frac{pq}{t}$, if t is very small.

When t is very small, the chords PQ and pq may be replaced by their arcs.

Therefore, if v be the velocity and a the acceleration at P,

$$\text{arc } PQ : \text{arc } pq :: v : a.$$

The velocity of the moving point in the hodograph is equal to the acceleration of the corresponding point in the path.

In tracing the form of the path it is immaterial what point is taken as the origin O. In fact we can pass from

O to O' as origin by compounding the vector $O'O$ with each vector from O.

The corresponding fact in the case of the hodograph is that the form of the hodograph is unaltered when a vector $o'o$ is compounded with each vector from o, i. e. when a constant velocity, which may be any whatever, is compounded with the velocity of the moving point.

Uniform circular motion. Let a point move uniformly with velocity v round a circle of radius r; then the vector which represents its velocity is of constant magnitude v, and its extremity traverses a circle with uniform velocity.

Let P, Q be two neighbouring points on the path, p, q the corresponding points on the hodograph.

Then if C and c are the centres of the circles the angles PCQ, pcq are equal.

Therefore $r : v :: \text{arc } PQ : \text{arc } pq$

$:: \quad v \quad : \quad a.$

Therefore the acceleration at P is $\frac{v^2}{r}$, and it is perpendicular to cp, that is, it is directed along PC.

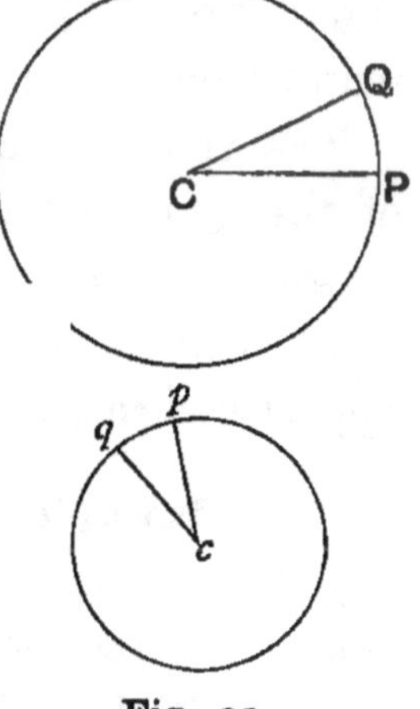

Fig. 21.

§ 12. **Rectilinear motion with uniform acceleration.**

If v_0 be the initial velocity, v the velocity acquired in a time t by a point which has uniform acceleration a, we have by definition

$$\frac{v - v_0}{t} = a,$$

or

$$v = v_0 + at. \tag{1}$$

Again, it has been shown that if s be the distance traversed,

$$s = \tfrac{1}{2}(v_0 + v)t. \tag{2}$$

Substituting in (2) the value of v given by (1) we have

$$s = v_0 t + \tfrac{1}{2} at^2. \qquad (3)$$

But we can also write (1) in the form

$$v_0 = v - at,$$

and substituting this value of v_0 in (2) we have

$$s = vt - \tfrac{1}{2} at^2. \qquad (4)$$

Again, by (1) $t = \dfrac{v - v_0}{a}$, and substituting this value of t in (2) we have

$$2as = v^2 - v_0^2. \qquad (5)$$

Each of the equations (1), (2), (3), (4), (5) contains a relation between four of the five quantities s, v_0, v, a, t.

When any three of these quantities are given, the other two can be deduced from two of the equations.

Thus if s, a, t are given, v_0 can be found directly from (3) and v from (4); and if v, v_0, and s are given, a can be found from (5) and t from (2).

The velocity and acceleration must be measured in terms of the units which are derived from the chosen units of length and time.

A velocity and an acceleration, which are denoted by opposite signs, have opposite directions.

Examples on uniformly accelerated motion.

1. The velocity of a point increases uniformly in 10 seconds from 150 to 200 centimetres per second. To find the acceleration and the distance traversed in the given time.

The initial velocity $= 150 = v_0$.
The final velocity $= 200 = v$.
The time of motion $= 10 = t$.

Let a be the acceleration, s the distance traversed.

Then $a = \dfrac{v - v_0}{t} = \dfrac{200 - 150}{10} = 5$.

And $s = \tfrac{1}{2}(v_0 + v)\, t = 5\,(150 + 200) = 1750$ centimetres.

2. In 5 seconds a body travels 125 cm. with acceleration 15. Find the initial velocity.

Here the distance s, acceleration a, and time t are given.

Equation (3) above gives the initial velocity v_0,

$$125 = 5\,v_0 + \tfrac{1}{2} \times 15 \times 25,$$

whence $$v_0 = -\tfrac{25}{2}.$$

The velocity is therefore initially in the direction opposite to that of the displacement (125 cm.) and of the acceleration. It is continually diminished and ultimately reversed.

To find when the velocity is zero, we write in equation (1) $v = 0$, $v_0 = -\frac{25}{2}$, $a = 15$, whence $t = \frac{5}{6}$ sec.

The distance traversed in this time is given by

$$s = \tfrac{1}{2} \times \left(-\tfrac{25}{2}\right) \times \tfrac{5}{6} = -\tfrac{125}{24}.$$

The negative sign indicates that this distance is in the direction opposite to that of the acceleration.

The body then retraces its path. It reaches its initial position at a time t given by writing $s = 0$ in equation (3) This gives

$$t = -\frac{2v_0}{a} = \tfrac{5}{3} \text{ sec.}$$

Hence the path of 125 cm. is really described as follows. In $\frac{5}{6}$ sec. the body moves $5\frac{5}{24}$ cm. in the direction opposite to the acceleration. In an equal succeeding interval it retraces the same path backwards, and then in $3\frac{1}{3}$ seconds it travels 125 centimetres with a constantly increasing velocity.

3. Find the initial velocity of a body which moving with acceleration 10 acquires in 5 seconds a velocity of 150.

4. Find the initial velocity of a body which moving with acceleration 15 acquires in 5 seconds a velocity 10.

5. Find the final velocity when the initial velocity is 100, the acceleration 18, and the time is (1) 4 seconds, (2) 6 seconds, (3) 8 seconds.

6. Substituting an acceleration -18 for 18, solve question 5.

7. A body after moving from rest with uniform acceleration for $\frac{1}{3}$ minute acquires a velocity of 10 metres per second. Find its acceleration.

8. A train moving at a speed of 60 kilometres per hour is pulled up in $\frac{1}{2}$ minute. Find its acceleration.

9. If the velocity of a moving body increase in 10 seconds at a uniform rate from 15 to 25, find the distance traversed during the given time.

10. If the distance traversed in 30 seconds is 15 metres and the final velocity is 35 cm. per sec., find the initial velocity.

11. In questions (3) and (4) find the distance traversed.

12. A body starts from rest, and in 5 seconds travels 100 cm. Find the acceleration.

13. If the acceleration of a body be 5, and in 5 seconds it travels 125 centimetres, find the initial velocity.

14. A body moves with uniform acceleration through 100 centimetres in 5 seconds and then comes to rest, find the acceleration, the initial velocity, and the average velocity during each second.

15. A body moves from rest with acceleration 32 foot-second units, find the distance traversed and the velocities acquired in 1, 2, 3 seconds.

Find also the average velocity during each second, and the distance travelled during the 10th second.

16. The velocity of a moving point increases during motion through 12 centimetres from 18 to 21 centimetres per second. Find the acceleration, and the time of motion.

17. In moving through 10 feet with acceleration 32 foot-second units a moving point has acquired a velocity of 26 feet per second. Find its initial velocity, and the time during which it has been moving.

18 A mass whose acceleration is uniform moves over 483 feet in the fifth second from rest. Find its acceleration, and its velocity at the beginning of the fifth second.

§ 13. Motion with Uniform Acceleration.

A particle starts from a point O with initial velocity u along Ox, and has acceleration g along Oy, a perpendicular to Ox. To find its motion.

The component of velocity along Ox is constant, and if $OM = MN = NP = u$, OM, ON, OP, are the horizontal distances travelled in 1, 2, 3 seconds respectively.

Also if $Om = \frac{1}{2}g$, $On = \frac{1}{2}g \cdot 2^2$, $Op = \frac{1}{2}g \cdot 3^2$; Om, On, Op are the vertical distances travelled in 1, 2, 3 seconds.

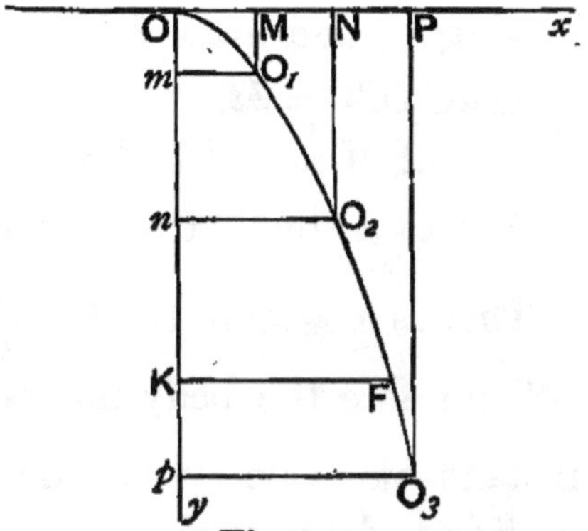

Fig. 22.

Completing the rectangles mM, nN, and pP, the actual positions of the particle after 1, 2, 3 seconds are respectively O_1 O_2 O_3; and the particle describes a continuous curve passing through these points.

Let F be the position of the particle after a time t, FK perpendicular to Oy.

Then $OK = \frac{1}{2}gt^2$, and $FK = ut$.

Therefore $$FK^2 = \frac{2u^2}{g} OK.$$

This relation enables us to find all points on the curve when u and g are known.

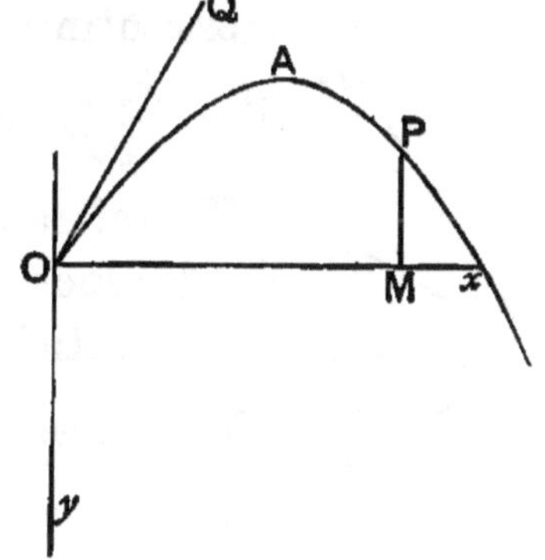

The curve thus determined is called a Parabola, and the quantity $\frac{2u^2}{g}$ is called its Parameter.

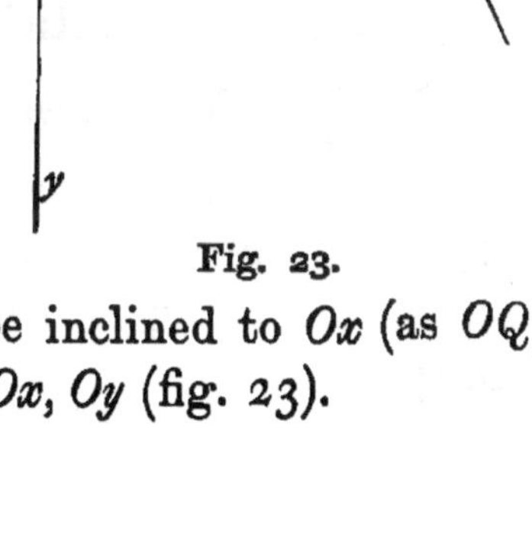

If the particle were projected from O_1, with a velocity compounded of u and g, it would describe the same parabola.

Fig. 23.

Next, let the initial velocity be inclined to Ox (as OQ), having components $h, -k$ along Ox, Oy (fig. 23).

Let P be the position of the particle at a time t after leaving O, and draw PM perpendicular to OM.

Then $OM = ht$, (1)

$PM = -kt + \frac{1}{2}gt^2$. (2)

And at a time t the velocity along Oy is $-k+gt$.

This is negative till $t = \frac{k}{g}$, and afterwards positive.

Therefore the body moves upwards until $t = \frac{k}{g}$; at this instant its velocity is horizontal, and afterwards the motion is downwards, the path being a parabola with parameter $\frac{2h^2}{g}$.

The point A is called the vertex of the parabola, and the vertical line through A is the axis.

This problem derives its interest from the fact that a falling body has uniform acceleration downwards, when the resistance of the air can be neglected. A falling body projected at any angle to the vertical is often called a Projectile, and its path a Trajectory.

The Projectile attains its greatest height when $t = \frac{k}{g}$; substituting this value of t in (2) we find that $\frac{k_2}{2g}$ is the greatest height attained.

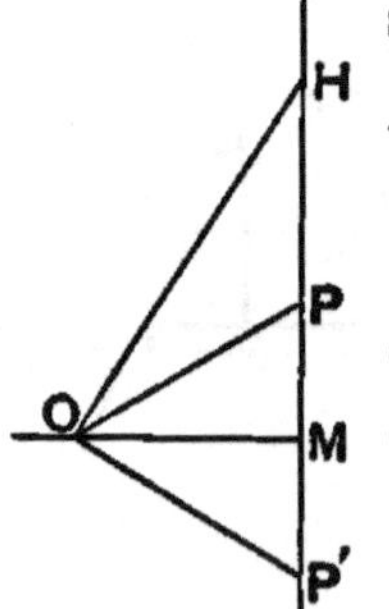

Fig. 24.

Therefore the height attained by the particle depends only on the vertical component of the initial velocity.

The Hodograph of uniformly accelerated motion.

Since the acceleration is constant in magnitude and direction, the hodograph is a straight line along which the tracing point moves with velocity g.

Let OH (fig. 24) represent the initial velocity in magnitude and direction, and $HP = gt$; then OP is the velocity at time t.

Let OP' and OP be equally inclined to the horizontal OM.

Then $OP = OP'$ and $MP = MP' = \frac{1}{2} PP'$.

Hence, considering two points of the path at which the tangents are equally inclined to the vertical, we see that the velocities at these points are equal in magnitude, and the time occupied in passing from one to the other is twice that occupied in passing from one of them to the vertex.

Hence since the horizontal velocity is constant, the two points are equally distant from the axis of the parabola.

Also, since the vertical velocities at the two points are the same, they are in the same horizontal line.

Hence the portions of the path on each side of the vertex A are precisely similar.

Examples on the motion of a Projectile.

In the following examples it may be assumed that the acceleration of a projectile is 980 centimetre-second units or 32 foot-second units.

1. A body is thrown vertically upwards with a velocity of 2058 centimetres per second. How high will it rise, and when will it strike the ground?

The initial velocity v_0 is 2058; the acceleration a is -980; the velocity v at the highest point is 0.

Therefore $v_0^2 + 2as = 0$.

And
$$s = \frac{v_0^2}{-2a} = \frac{2058 \times 2058}{2 \times 980} = 2160.9 \text{ cm.}$$

Also if t be the time to the highest point

$$t = \frac{s}{\frac{1}{2}(v_0 + v)} = \frac{2160.9}{1029} = 2.1 \text{ secs.}$$

This is also the time of fall from the highest point.

Therefore 4·2 seconds is the time of flight.

2. A projectile is fired at an inclination 60° above the horizon, with a velocity of 588 centimetres per second. Find the horizontal and vertical components of velocity after 3 seconds.

The horizontal component of velocity is initially $588 \times \frac{1}{2}$ or 294. Since there is no horizontal acceleration, this is the horizontal component of velocity at any later time.

The vertical component of velocity v_0 is initially $294\sqrt{3}$, and the vertical component after 3 seconds is

$$v_0 - gt \text{ or } 294\sqrt{3} - 3 \times 980 \text{ upwards};$$

i. e. $294\,(10 - \sqrt{3})$ downwards.

3. A body is projected in a horizontal direction, with a velocity of 14 metres per second, from the top of a tower 22.5 metres high.

(*a*) Find the distance from the foot of the tower at which the body will strike the ground.

Let t be the time of fall.

The vertical component of velocity is initially zero, and the vertical distance fallen is 2250 centimetres.

Therefore $2250 = \frac{1}{2} \times 980\, t^2$,

$$\text{or} \quad t = \frac{\sqrt{225}}{\sqrt{49}} = \frac{15}{7}.$$

The horizontal velocity is initially 1400, and is uniform.

Therefore the required distance is $1400 \times \frac{15}{7}$ or 3000 centimetres.

(*b*) Find the velocity of the body when it touches the ground.

The horizontal component is 1400.

Since the vertical component is initially zero, its final value is given by $v = gt = \frac{15}{7} \times 980 = 2100$.

Therefore the resultant velocity is

$$\sqrt{(1400)^2 + (2100)^2} \text{ or } 700\sqrt{13}.$$

The direction is inclined below the horizon at an angle a, such that

$$\tan a = \tfrac{3}{2}.$$

4. Three seconds after a stone has begun to fall down the shaft of a mine, a second stone is thrown down the shaft with a velocity of 4410 centimetres per second, and both stones strike the bottom simultaneously. Find the time of fall of the first stone and the depth of the shaft.

5. A ball projected vertically upwards rises 102.9 metres in

3 seconds. Find, (1) how much longer, (2) how much higher the ball will rise, (3) its velocity at the instant named.

6. Two bodies are simultaneously projected upwards in directions which make angles 30° and 60° respectively with the vertical. Prove that if they both rise to the same height, they strike the ground at the same instant, but one will be three times as far from the point of projection as the other.

7. A body is thrown vertically downwards with an initial velocity of 48 feet per second. Find the distance traversed in the fourth second.

8. Two particles are simultaneously thrown vertically upwards from the same point. One has an initial velocity of 144 feet per second, and the other a velocity of 202 feet per second. Find the height of the latter when the former reaches the ground again.

9. A cricketer throws a ball 48 yards. If it rises 36 feet in the air, find its initial velocity and the time of flight.

10. A ball is thrown upwards at an angle 60° with the horizon. If its initial velocity is 48 feet per second, find when it hits the ground.

11. Show that if a ball is projected with an initial velocity of given magnitude, its range on a horizontal plane will be greatest when the direction of the initial velocity is inclined at 45° to the horizon.

12. A ball is projected horizontally from a railway carriage in a direction perpendicular to that in which the train is travelling.

If the speed of the train be 30 miles an hour, the initial velocity of the ball 33 feet per second relatively to the thrower, and the height of the point of projection above the line 9 feet, find the velocity of the ball when it hits the ground, and the horizontal distance which it travels.

§ **14.** The following discussion of the motion of a projectile may be useful. It is assumed that the reader is acquainted with the principal properties of the parabola.

Let v be the initial velocity, a the angle of projection, i.e. the angle between v and the horizontal.

Then $v \cos a$, $v \sin a$ are the horizontal and vertical components of the initial velocity, the axis of y being drawn upwards from O.

And if x, y are the horizontal and vertical components of displacement in time t,

$$x = vt \cos a,$$

$$y = vt \sin a - \tfrac{1}{2} gt^2.$$

Eliminating a,

$$x^2 + (y + \tfrac{1}{2} gt^2)^2 = v^2 t^2.$$

Therefore if several particles are projected simultaneously in the same plane from a point, with velocities of equal magnitude, they all lie at a time t on a circle with radius vt, and centre $(0, -\tfrac{1}{2} gt^2)$.

That is, the centre of the circle descends with acceleration g, and the radius increases with velocity v.

Again, eliminate t.

$$\text{Then} \quad y = x \tan a - \tfrac{1}{2} \frac{gx^2}{v^2 \cos^2 a},$$

$$\text{or} \quad \tfrac{1}{2} gx^2 (1 + \tan^2 a) - v^2 x \tan a + v^2 y = 0. \qquad (1)$$

It has already been shown (§ 13) that this is a parabola, with its axis vertical and its latus rectum equal to

$$\frac{2v^2 \cos^2 a}{g}.$$

If in § 13 Ox be inclined to the horizon, and FK be drawn parallel to Ox, it can be easily shown, by compounding the uniformly increasing displacement along Ox with the uniformly accelerated displacement downward, that $FK^2 = \frac{2v^2}{g} OK$, v being the velocity of projection.

Hence the focus of the parabola is distant $\frac{v^2}{2g}$ from the point of projection, and since the tangent at O is equally inclined to the axis and to the focal distance, the position of the focus is completely known; for it is the point $\left(\frac{v^2}{2g}\sin 2\alpha, -\frac{v^2}{2g}\cos 2\alpha\right)$.

Since $\frac{v^2}{2g}$ does not involve α, the foci of all trajectories described from O with a given initial velocity lie on a circle with centre O and radius $\frac{v^2}{2g}$.

In equation (1) let x and y be known; then the equation is a quadratic in $\tan\alpha$. Hence there are generally two paths by which a projectile starting with an initial velocity v can reach a given point, and if α_1, α_2 are the values of α which satisfy the equation (1),

$$\tan\alpha_1 + \tan\alpha_2 = \frac{2v^2}{gx}.$$

The roots of the equation are imaginary if $v^2 < 2gy$, and this obviously is the case, for the given point cannot be reached.

To find the range of a projectile on an inclined plane through the point of projection, when the direction of projection and a normal to the plane lie in the same vertical plane.

Let β be the inclination of the plane.

Then it is required to satisfy equation (1) when

$$y = x\tan\beta.$$

Substituting in (1) and dividing by x,

$$\tan\beta = \tan\alpha - \tfrac{1}{2}\,\frac{gx}{v^2\cos^2\alpha}:$$

$$\text{or} \qquad x = \frac{2v^2}{g}\;\frac{\sin(\alpha-\beta)\cos\alpha}{\cos\beta};$$

and the range (i.e. the distance travelled on the plane) is

$$x\sec\beta, \quad \text{or} \quad \frac{2v^2\sin(\alpha-\beta)\cos\alpha}{g\cos^2\beta},$$

$$\text{or} \quad \frac{v^2[\sin(2\alpha-\beta)-\sin\beta]}{g\cos^2\beta}.$$

Now β is fixed. Therefore the range is greatest when $\sin(2\alpha-\beta)=1$, that is, when $2\alpha-\beta=\frac{\pi}{2}$, i.e. when the direction of projection bisects the angle between the vertical and the line of greatest slope on the plane.

The greatest range is $\frac{v^2}{g(1+\sin\beta)}$, and is greater on a horizontal plane than on any other, for in this case $\beta=0$.

§ 15. Angular Acceleration, about a fixed axis.

The angular acceleration of a body is the rate of increase of its angular velocity. We cannot here consider the general case when the axis of rotation changes during motion. We shall only take the simple case in which the axis is fixed and the rate of rotation alters.

If ω_0 and ω be the values of the angular velocity at the beginning and end of the time t,

$\frac{\omega-\omega_0}{t}$ is the angular velocity gained in unit time.

Therefore, a being the angular acceleration,

$$\omega-\omega_0=at. \tag{1}$$

Also we have already shown in §10 that

$$\frac{\omega+\omega_0}{2}t=d. \tag{2}$$

Multiplying (1) and (2) together,

$$\omega^2-\omega_0{}^2=2ad.$$

§ 16. Composition and Resolution of Accelerations.

Let PQ represent the acceleration of a moving point, OP its velocity at a given instant; then OQ is the velocity after unit time.

Take any two lines Ox, Oy through O in the plane OPQ, and resolve OP and OQ into their components OE, EP and OF, FQ along these lines.

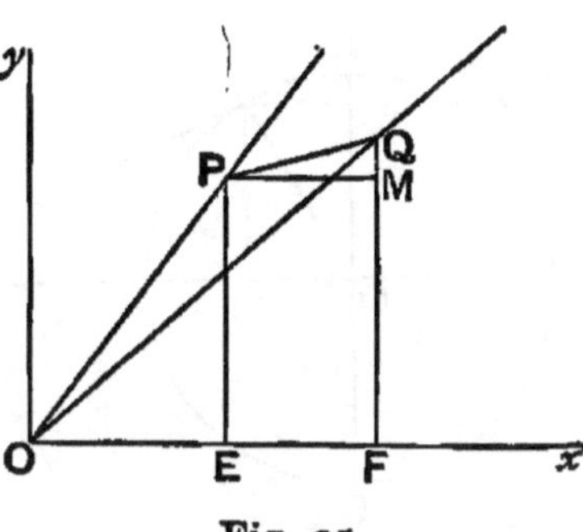

Fig. 25.

Draw PM parallel to EF.

Then EF or PM is the gain of velocity parallel to Ox in unit time, and MQ is the gain of velocity parallel to Oy in unit time, and by compounding these we get the acceleration PQ.

Therefore the resultant acceleration is obtained by compounding the rates of increase of the component velocities parallel to Ox and Oy; these rates of increase are called components of acceleration.

The most useful case is that in which xOy is a right angle.

Conversely, an acceleration can be resolved into its components along any two lines co-planar with it.

Relative Acceleration.

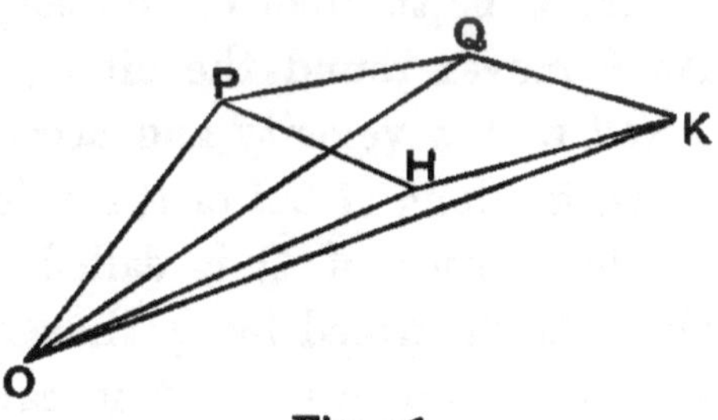

Fig. 26.

Let OP, OQ denote the initial velocities of two moving points A and B; PH, QK their accelerations.

Then OH, OK are the velocities after unit time, and the velocity of B relatively to A is initially PQ and finally HK.

Therefore the relative acceleration, or change of relative velocity, is found by compounding HK and QP.

And since the resultant of HK, KQ, QP, PH is zero, QP and HK are equivalent to QK and HP.

Therefore the acceleration of B relatively to A is obtained by compounding the acceleration of B and the reversed acceleration of A, i.e. it is the vectorial difference of the accelerations of B and A.

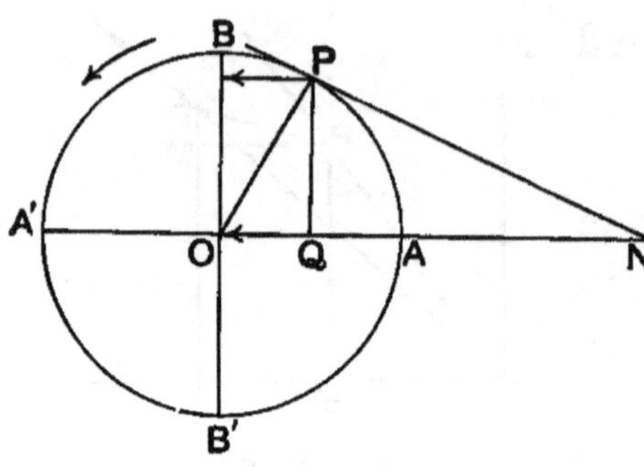

Fig. 27.

§ 17. Simple Harmonic Motion.

Let APA' be a circle of radius r, AOA', BOB' perpendicular diameters, P any point on the circle.

Then if P moves round the circle in the direction of the arrow with uniform angular velocity ω, its velocity is $r\omega$ directed along NP the tangent at P, and its acceleration is $r\omega^2$ directed along PO.

The time occupied by the moving point in going round the circle is $\frac{2\pi}{\omega}$.

Let Q be the foot of the perpendicular from P on AOA'. As P moves round the circle, Q moves to and fro along $A'OA$ with a velocity and acceleration which are the components along $A'OA$ of the velocity and acceleration of P.

The motion of Q is called a simple harmonic motion; the path described by Q when P goes once round the circle starting from any point is called a complete oscillation; the time $\frac{2\pi}{\omega}$ occupied in describing this path is called the time of a complete oscillation or the periodic time; the distance OA is called the amplitude of vibration.

Let $OQ = x$. Then $PQ^2 = r^2 - x^2$.

The component of the velocity of P along AA' is $r\omega \frac{QN}{PN}$; this is equal to $r\omega \frac{PQ}{OP}$ or $\omega\sqrt{r^2 - x^2}$.

The acceleration of Q is of magnitude $r\omega^2 \frac{x}{r}$, or $\omega^2 x$; and it is always directed towards O: it is therefore completely represented by $-\omega^2 x$.

Thus in a simple harmonic motion the acceleration is directed to a fixed point in the line of motion, and is proportional to the distance x from the point.

Since the time of vibration is $\frac{2\pi}{\omega}$, it depends only on the law of acceleration, i.e. on the constant ω, and not on the amplitude.

Phase. The fraction of the time of a vibration which has elapsed in the passage of the moving point from A to Q is called the phase of the motion at Q.

It is $\frac{\theta}{2\pi}$, where θ is the angle AOP.

Examples.

1. Two particles are projected upwards with the same velocity from two points A and B in the same vertical, B being at a height h above A. If the second particle be projected t seconds after the first, and if u be the velocity of the first particle when it meets the second, then

$$h = ut + \tfrac{1}{2}gt^2.$$

2. A ball falling from the top of a tower has descended a feet when another ball is let fall from a distance b below the top. Show that if they reach the ground together, the height of the tower is $\frac{(a+b)^2}{4a}$ feet.

3. A bird wishes to reach a point due N. when the wind is blowing from the S.W. at 20 miles an hour. Given the velocity with which the bird can fly, find the point of the compass for which it must aim.

Show that if it goes less than 14.1 miles per hour, it cannot reach the point.

4. A railway train starts from the top of a straight incline 1·1 miles long with a velocity of 22 miles per hour, and on reaching the bottom has acquired a velocity of 32 miles per hour. Find the acceleration in foot-second units.

5. An engine-driver suddenly puts on his break and shuts off the steam when the train is running at full speed. In the first second afterwards the train travels 87 feet, and in the next 85 feet. Find the original speed of the train, the time which it will take to come to rest, and the distance that it will travel in the interval, assuming the break to cause a constant retardation.

Find also the time which the train would take, if it were 96 yards long, to pass a spectator standing 484 yards ahead of the train at the moment when the break was applied.

6. A body moves in 2 seconds through a distance of 100 feet from rest, its acceleration being constant. Find the distance traversed in the next second.

7. A bullet is fired with a velocity of 1920 feet per second in a direction equally inclined to the horizontal and the vertical; find the height to which it will rise, and the velocity at the highest point.

8. Assuming that the earth is a sphere of 4000 miles radius turning on its axis in 86164 seconds, find its angular velocity, and the acceleration of a point on its surface—(1) at the equator, (2) in latitude 45°, the foot and second being taken as units.

9. The hour and minute-hands of a clock are respectively 2 inches and 3 inches long. Find the accelerations of their extremities in foot-second units.

Find also the acceleration of the extremity of the hour-hand relatively to that of the minute-hand at 12.0, at 6.0, at 3.0, and at 4.30.

Appendix.

Proofs of some simple geometrical results are appended for the convenience of beginners.

(*a*) *The circular measure of an angle.*

If r is the radius of a circle, the circumference is $2\pi r$,

π being the incommensurable number $3 \cdot 14159 \ldots$, or $3\frac{1}{7}$ approximately.

Let O be the centre of a circle ABC (fig. 28).

Since angles at the centre are proportional to the arcs on which they stand,

$$\frac{\text{angle } AOB}{\text{four right angles}} = \frac{\text{arc } AB}{\text{whole circumference.}}$$

But the circumference is $2\pi r$.

Therefore $$\frac{\text{angle } AOB}{\text{four right angles}} = \frac{\text{arc } AB}{2\pi r}. \qquad (1)$$

Let the unit angle be chosen so that 2π is the measure of four right angles.

This unit is $\dfrac{180^\circ}{3 \cdot 14159 \ldots}$ or $57^\circ \cdot 295779 \ldots$.

This is called the **Radian**, and the measure of an angle in terms of radians is called its circular measure.

By (1), the angle AOB is $\dfrac{AB}{r}$ radians or simply $\dfrac{AB}{r}$, if it is understood that the radian is the unit.

The circular measure of an angle at O is therefore the

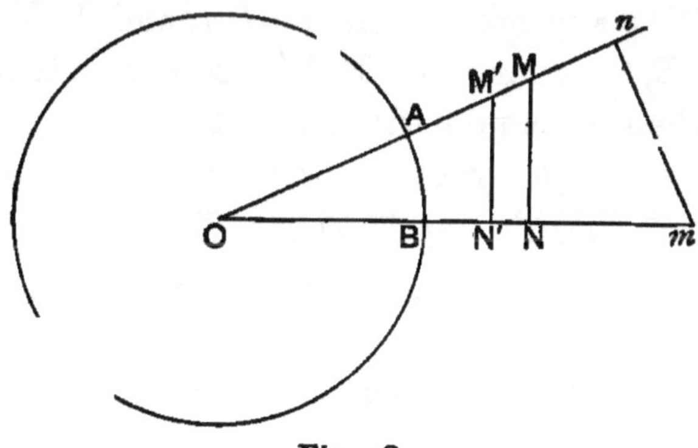

Fig. 28.

arc which it subtends on a circle of unit radius with O as centre. A right angle is $\frac{\pi}{2}$, i.e. it contains $\frac{\pi}{2}$ Radians.

(*b*) ***Definitions of the sine, cosine, and tangent of an angle.***

Let AOB be any angle less than $\frac{\pi}{2}$.

From M on OA draw MN perpendicular to OB. Then in the triangle MON, MN opposite to the angle AOB is called the perpendicular, ON adjacent to AOB is called the base, and OM is the hypotenuse.

The ratio $\dfrac{\text{perpendicular}}{\text{hypotenuse}}$ or $\dfrac{MN}{OM}$ is called the sine of AOB,

. . . . $\dfrac{\text{base}}{\text{hypotenuse}}$ or $\dfrac{ON}{OM}$ is the cosine of AOB,

. . . . $\dfrac{\text{perpendicular}}{\text{base}}$ or $\dfrac{MN}{ON}$ is the tangent of AOB.

For brevity these ratios are denoted by

$$\sin AOB, \quad \cos AOB, \quad \tan AOB.$$

These ratios and their reciprocals are called the trigonometrical ratios of the angle AOB.

Their magnitudes do not depend on the position of M on OA.

For if M' be any other point on OA, and $M'N'$ perpendicular to OB, the triangles $OM'N'$, OMN have their angles equal and their sides proportional, so that

$$\frac{MN}{OM} = \frac{M'N'}{OM'}, \quad \frac{ON}{OM} = \frac{ON'}{OM'}, \quad \frac{MN}{ON} = \frac{M'N'}{ON'}.$$

Again, take m on OB, and draw mn perpendicular to OA.

The triangle mOn, MON have their angles equal, each to each, and $\dfrac{MN}{OM} = \dfrac{mn}{Om}$, $\dfrac{ON}{OM} = \dfrac{On}{Om}$, $\dfrac{MN}{ON} = \dfrac{mn}{On}$.

Thus the values of the trigonometrical ratios deduced from the triangles Omn, OMN are the same, and M may be taken on either of the two lines which form the angle at any distance from O; the trigonometrical ratios of the angle AOB will be the same for all positions of M.

The ratios are different for different angles; and when the ratios of an angle are known, the angle can be found.

(c) *By diminishing the angle AOB, its cosine may be made to differ from 1 by less than any assignable quantity.*

Let ABD be a circle with centre O, and diameter BD.

Draw AM perpendicular to BD, and join AD, AB.

Then $$1-\cos AOB = 1-\frac{OM}{OA} = 1-\frac{OM}{OB}$$
$$=2.\frac{MB}{BD}=2.\frac{MB}{BA}.\frac{BA}{BD}.$$

But the triangles AMB, DAB have their angles equal, each to each.

Therefore $$\frac{MB}{BA}=\frac{BA}{BD}.$$

Therefore $$1-\cos AOB = 2.\frac{BA^2}{BD^2}.$$

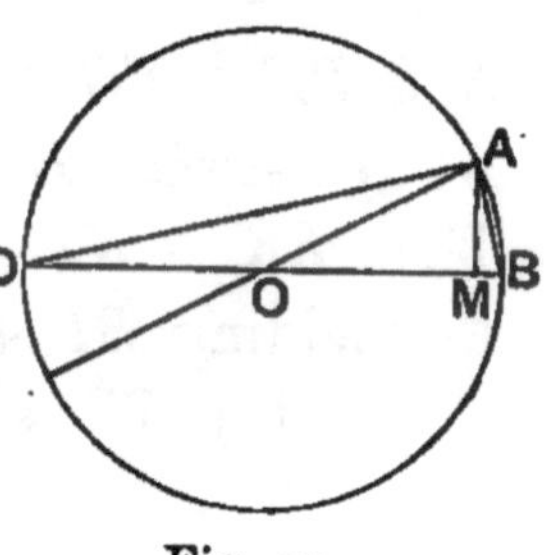

Fig. 29.

Now by revolving OA round O towards OB the arc AB can be made smaller than any given fraction of the diameter BD.

Therefore $\frac{BA}{BD}$ and *à fortiori* $\frac{BA^2}{BD^2}$ can be made less than any assignable quantity by diminishing the angle AOB.

Therefore $\cos AOB$ may be made to differ from 1 by a quantity less than any assignable quantity.

Since $\frac{BA}{BD} = \sin ADB = \sin \frac{1}{2} AOB$, we have

$$1-\cos AOB = 2\,(\sin \tfrac{1}{2} AOB)^2, \text{ or } 2\sin^2 \tfrac{1}{2} AOB.$$

A tangent to a curve is defined as follows—

Let P and Q be two points on a curve, PQ the chord

joining them. Then if P remains fixed, and Q moves along the curve towards P, the line along which PQ lies when Q coincides with P is called the tangent at P.

(*d*) *By diminishing the arc BC of a curve, it may be made to differ from its chord by less than any fraction of the chord, however small.*

Draw the tangents at B and C to the arc and let them intersect in D.

Then the arc is greater than its chord; it is also less than the sum of the distances BD, DC.

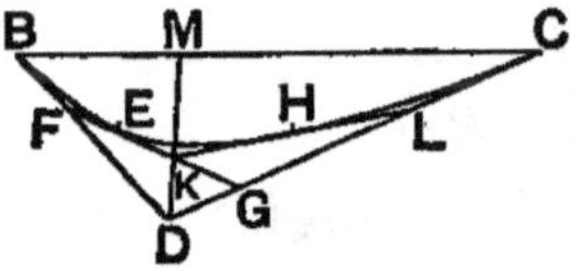

Fig. 30.

For at a point E in BC, draw a tangent FEG.

Then $FG < FD + DG$.

And $BF + FG + GC < BD + DC$.

At H between E and C draw a tangent KHL.

Then $EK + KL + LC < EG + GC$.

And adding $BF + FE$ to each, we have

$$BF + FK + KL + LC < BF + FG + GC$$
$$< BD + DC \text{ à fortiori.}$$

Dividing the arcs between B and C any number of times, we find that the broken line between B and C formed by the tangent constantly diminishes in length as the number of tangents increases, and at the same time tends to coincide more and more closely with the curve.

Therefore the length of the arc BC is less than the sum of BD, DC and greater than BC.

From D draw DM perpendicular to BC.

$$\text{Then}\quad \frac{BM}{BD} = \cos DBM,$$
$$\frac{MC}{CD} = \cos DCM.$$

Now by diminishing the arc BC we can make $\cos DBM$, $\cos DCM$ each differ from unity by less than any assignable quantity.

Let $\cos DBM = 1-h$, $\cos DCM = 1-k$.

$$\text{Then} \quad BM = (1-h)\, BD.$$
$$MC = (1-k)\, CD.$$
$$\text{Therefore} \quad BC = BD + CD - (h.\, BD + k.\, CD).$$

But h and k can be made smaller than any assigned quantity by diminishing BC.

Therefore $\dfrac{BD+CD}{BC}$ differs from 1 by $h\dfrac{BD}{BC} + k\dfrac{CD}{BC}$, which can be made less than any assigned quantity by diminishing BC.

Therefore, since $BD + CD <$ arc BC, $\dfrac{\text{arc } BC}{\text{chord } BC} - 1$ can be made less than any assignable quantity by diminishing BC.

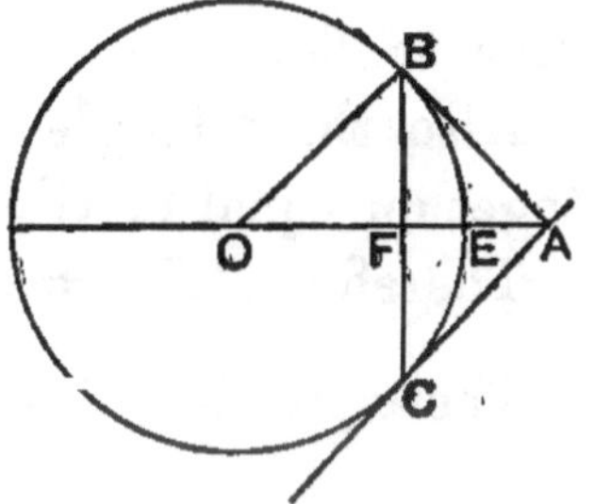

Fig. 31.

Example. Let the arc BC be a portion of a circle with centre O, and let the tangents at B and C intersect on OA.

Then the arc BE lies between BF and BA in magnitude, for it is half of the arc BC.

Now $AB^2 = AF^2 + FB^2$,

$$\text{or} \quad 1 = \frac{AF^2}{AB^2} + \frac{FB^2}{AB^2};$$

but since FAB, BAO are similar triangles,

$$\frac{AF}{AB} = \frac{AB}{AO}.$$

$$\text{Therefore} \quad 1 = \frac{AB^2}{AO^2} + \frac{FB^2}{AB^2}.$$

Thus $\dfrac{FB}{AB}$ can be made to differ from 1 by a fraction smaller than any assigned, if BC is made very small.

E.g. Let $\dfrac{AB}{OB} = \dfrac{1}{100}$, then $\dfrac{BF^2}{AB^2}$ differs from 1 by $\dfrac{1}{10000}$.

But if $\frac{AB}{OB} = \frac{1}{1000}$, $\frac{BF^2}{AB^2}$ differs from 1 by $\frac{1}{1000000}$.

And $\frac{BF}{AB}$ is nearer to 1 than is $\frac{BF^2}{AB^2}$

Thus, to whatever degree of accuracy calculations are made, it is possible to reduce the arc BC to so small a magnitude that its ratio to the chord BC differs from 1 by a fraction insensible in any calculations.

This is expressed by saying that the arc of a curve when indefinitely diminished is equal to its chord.

(*e*) *To find the values of the sine, cosine, and tangent of* 30°, 45°, *and* 60°.

Let ABC be an equilateral triangle, D the middle point of its base. Then it is easy to show that AD is perpendicular to BC.

Also the triangle is equiangular, and all its angles are together equal to 180°.

Therefore $ABD = 60°$, and $BAD = 30°$.

Therefore $\sin 60° = \frac{AD}{AB} = \cos 30°$.

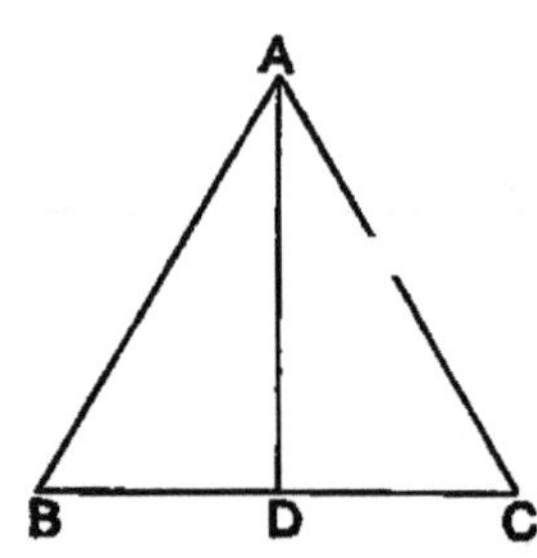

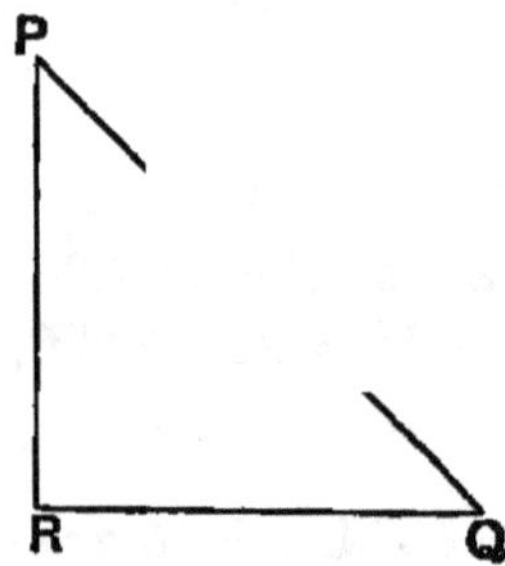

Fig. 32.

$$\cos 60° = \frac{BD}{AB} = \sin 30°.$$

Now $BD = \frac{1}{2} BC = \frac{1}{2} BA$.

$$\text{Therefore} \quad \frac{BD}{AB} = \frac{1}{2}.$$

Also $AD^2 + BD^2 = AB^2$.

But $BD^2 = \frac{1}{4} AB^2$;

therefore $AD^2 = \frac{3}{4} AB^2$,

and $\dfrac{AD}{AB} = \dfrac{\sqrt{3}}{2}$.

Therefore $\sin 60° = \cos 30° = \dfrac{\sqrt{3}}{2}$,

and $\cos 60° = \sin 30° = \frac{1}{2}$,

$$\tan 60° = \frac{\sin 60°}{\cos 60°} = \sqrt{3},$$

$$\tan 30° = \frac{\sin 30°}{\cos 30°} = \frac{1}{\sqrt{3}}.$$

Next, let PQR be an isosceles triangle right-angled at R. Then RPQ, RQP are equal angles, and each of them is 45°. Also $PQ^2 = PR^2 + RQ^2 = 2\,PR^2$ or $2\,RQ^2$.

Therefore $\sin PQR = \sin 45° = \dfrac{PR}{PQ} = \dfrac{1}{\sqrt{2}} = \dfrac{RQ}{PQ} = \cos 45°$,

$$\tan 45° = 1.$$

CHAPTER II.

THE LAWS OF MOTION.

THE Laws of Motion were first stated by Newton; they are deductions from observation and experiment, and their truth cannot be proved mathematically.

They stand in the same relation to the Science of Mechanics as the Definitions and Axioms of Euclid to the Science of Geometry.

We shall for the present consider all parts of a body as having the same velocity; our conclusions will only apply to particles or to bodies which can be regarded as particles.

§ 1. **Law I.** *Every body continues in a state of rest or of uniform motion in a straight line, except in so far as it is compelled to change that state by external force acting upon it.*

The following assertions are implied in this Law:—

A material body possesses Inertia, that is, its state of motion cannot of itself change. If a change takes place it is due to a cause external to the body; this cause is called Force.

Motion with uniform velocity (i.e. velocity of constant magnitude and direction) resembles rest in remaining unchanged except by the action of an external cause. Any change in the direction or magnitude of the velocity *implies** change of motion.

If a body moves with uniform velocity, the resultant of the forces which act on it is zero.

§ 2. Measurement of Time.

If a moving body is not acted on by external forces, it can be used as a time-keeper, for those times are equal in which the body moves over equal distances.

The difficulty is to ascertain whether external causes affect the motion. Suppose, however, that there are several bodies whose motions, if affected by external causes at all, are probably affected by different causes, or by the same causes to very different extents. Suppose further that the velocities of the bodies are either the same or are in a constant ratio to each other. There is then a high probability that the velocity of each body is uniform, and observations of the motion of one body in the system would afford a means of measuring time.

These are the ideas on which the experimental measurement of time is based, but the phenomena are not quite so simple, for no permanently uniform rectilinear motions are known. The standard motion is the rotation of the earth on its axis, times being taken as equal in which the earth revolves through equal angles. Now there is no reason *à priori* for regarding the earth's rotation as uniform. It is certainly retarded to a greater or less extent by the friction of the tides. It is also retarded if the earth moves in a resisting medium, or if the mass of the earth is augmented by showers of meteoric dust. On the other

* We do not say that change of velocity *is* change of motion; for it will be seen later that the mass of a body as well as the velocity must be considered in estimating change of motion.

hand, the earth is cooling, and its rotation may be accelerated by the consequent contraction.

Thus arises the difficulty stated above; do these causes sensibly affect the rate of rotation?

Let us consider the consequences of supposing that the earth does not rotate uniformly, but that its angular velocity is diminishing.

The effect of this would be to gradually lengthen the day; other time-keepers, such as watches and clocks, would constantly gain. Physical constants, such as the velocity of light and of sound, which involve time, would be affected to a constantly increasing extent. But much the most delicate test would be afforded by astronomical phenomena; the hypothesis that the earth rotates uniformly is in good accordance with the observed facts, and it seems impossible to reduce these facts to an organised whole on any other hypothesis.

We therefore adopt the hypothesis that the earth rotates uniformly; there is, however, a small discrepancy in the moon's motion which could be accounted for by supposing that the earth loses twenty-two seconds of time in a century, through retardation by tidal friction. So small an error in a time-keeper is quite insignificant.

The validity of the theory of Mechanics does not depend on the existence of a perfect time-keeper, but its results cannot be experimentally confirmed unless we have a good method of measuring time.

§ 3. **Law II.** *Change of Motion is proportional to the impressed force, and takes place in the direction in which the force acts.*

Consider the force exerted on a given particle during a given time τ. The change of motion then consists in the change of the velocity of the particle; and if OA, OB respectively denote its velocity at the beginning and end of the time τ, AB denotes the velocity gained in this time (Chap. I, § 11).

Since the law says nothing as to the initial state of rest or motion, we conclude that the change of velocity does not depend on this state; thus, if the initial velocity be $O'A$ and the same force act for the same time, the final velocity is $O'B$.

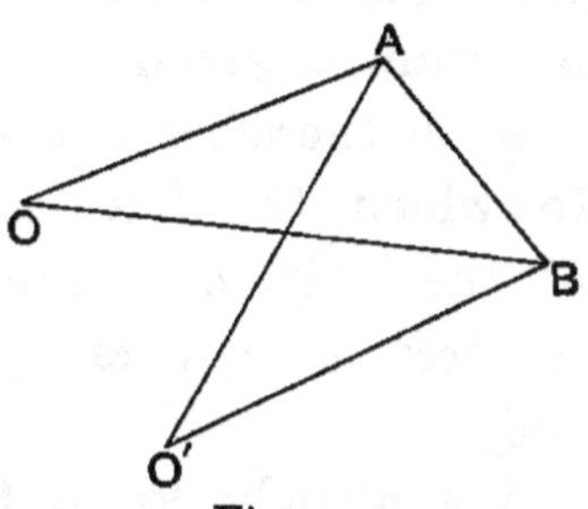

Fig. 33.

If the force be doubled, the increase of velocity is doubled; and as the velocity gained is in the direction of the force, force must be regarded as a vector, for it is only completely defined when its magnitude and direction are known.

Let v denote the velocity AB, communicated to the particle by the action of a force P during the time τ. If the same force continues to act during a second interval τ, it produces an additional velocity v, and the velocity generated in a time 2τ is $2v$. Thus the velocity produced by a force is proportional not only to the force but also to the time during which the force acts.

Let P, P' be different forces; v, v' the velocities they generate in times t, t' respectively, when acting on the same body.

Then $Pt : v :: P't' : v'$.

Denote the ratio $\dfrac{Pt}{v}$ or $\dfrac{P't'}{v'}$ by m.

Then m does not depend on the force, or on the time during which it acts, or on the velocity communicated to the body.

The product Pt is called the Impulse of the force P for the time t.

§ 4. Definition of Mass.

It is a matter of common experience that equal forces acting on different bodies for equal times do not impart to them equal velocities. The greater the load that a given engine has to draw, the longer is the time required to attain a given speed.

m is therefore a constant, which depends on the body to which the force is applied. It is called the Mass of the body, or, since it is proportional to the impulse required to generate unit velocity, the Inertia of the body.

We already know that the mass of a body does not depend on the acting force or the rate of motion; and physical experiments prove that mass is a definite physical quality of a body, independent (as far as we can tell) of its physical state—whether it be heated, electrified, magnetised or strained—and of the condition of neighbouring bodies.

It is therefore convenient to regard mass as a fundamental physical quantity.

The scientific unit of mass is the gram. It is $\frac{1}{1000}$th part of a piece of platinum in the Archives at Paris, and is approximately the mass of 1 ccm. of water at its maximum density.

The British unit of mass is the Pound Avoirdupois, containing 7000 grains or 453·5926 .... grams.

The gram contains 15·4323 .... grains.

The units of mass, length, and time being defined, all other units can be derived from these.

The system based on the centimetre, gram, and second as units is called the centimetre-gram-second system, or for brevity the C. G. S. system; that based on the foot, pound, and second is called the foot-pound-second system.

Density.

The Density of a body is measured by the mass contained in the unit of volume.

If M be the mass of the body and V its volume, its density is $\frac{M}{V}$. Denoting this by ρ, we have $M = \rho V$.

Making $M = 1$ and $V = 1$, we find that the unit of density is the density of a body whose unit of volume contains the unit of mass. This is therefore a derived unit.

Thus in the C. G. S. system 1 ccm. of the body whose density is 1 has a mass of 1 gram.

Since bodies alter in size when heated or compressed, the density of a body depends on its temperature and pressure.

Under ordinary atmospheric pressure water contracts when it is heated from 0° C (the freezing-point) to about 4° C, and expands when heated above 4° C. The density of water is therefore a maximum when its temperature is about 4° C, and it is then very nearly 1, being actually 1·000013.

Since the mass of a cubic foot of water is about 1000 oz. or 62½ lbs., the density of water is 62½ in the foot-pound-second system of units.

§ 5. Momentum.

If a mass m move with velocity v, the product mv is called the Momentum of the mass. Momentum, like velocity and displacement, is defined when its direction and magnitude are given. The unit of momentum is the momentum of unit mass moving with unit velocity.

In the relation $Pt = mv$, make mv and t each equal to 1.

$$\text{Then } P = 1.$$

Therefore the unit of force is that which generates unit momentum in unit time.

If the velocity increases uniformly during the time t, $\frac{v}{t} = a$ and $P = ma$. Making m and a each equal to 1, we find that the unit force produces unit acceleration in unit mass. When the acceleration is variable the time t must be made indefinitely short.

The unit of force in the C. G. S. system is called the dyne; in the foot-pound-second system it is called the poundal.

We may now state the Second Law of Motion in more definite terms.

The product of the quantities which represent the acceleration and mass of a moving body is proportional to the acting force; and the force and acceleration are in the same direction.

The relation $Pt = mv$, which also expresses the Law, amounts to the statement that—

The impulse of a force is equal to the momentum generated and is in the same direction.

Impulsive Forces.

Certain forces act for so short a time that it is impossible to find the acceleration, or the precise time of their action. We can only compare the velocities and thence the momenta of the moving bodies before and after action.

A simple instance is that of the collision of two billiard balls, or of the impact of a particle on a fixed plane.

Let a particle of mass m impinge perpendicularly on a fixed plane with velocity v, and rebound with velocity v'.

The total change of momentum is $-m(v+v')$, and this is the impulse of the force which the plane has exerted on the particle.

Forces of this kind are called Impulsive Forces. There is no essential difference between them and others, but the extreme brevity of the time during which they act makes it convenient to measure them by their impulses.

§ 6. Weight.

All bodies with which we are acquainted tend to fall to the earth, and if not restrained by some support, they fall with constantly increasing velocity. This must be due to a downward force measured by the product of the falling mass and its acceleration. This force is called the Weight of the body, or (for reasons which will appear later) the force with which the earth attracts it.

It was demonstrated first by Galileo, and afterwards by Newton, that at the same place all bodies, when their motion is not affected by the resistance of the air, fall from rest to the earth from the same height in the same time, whence it follows that the acceleration of all freely falling bodies is the same in the same place. Accurate experiments with the pendulum (to be explained later) have shown that, in different latitudes or at different heights above sea-level, the acceleration of a falling body is not the same.

The weight of a body consequently depends on the place in which it is measured; but the weights of different bodies at the same place are in the ratio of their masses, whatever the place may be.

It is important to distinguish clearly between the mass and the weight of a body.

The mass is measured by the impulse required to produce unit velocity, and is the same whatever be the position of the mass relatively to the earth or other bodies. The weight is the force with which the earth apparently attracts the body, and it varies with the position of the body.

The acceleration of a freely falling body in London is about 32·2 foot-second units or 981 centimetre-second units. It is generally denoted by g.

Hence the poundal is about $\frac{1}{32.2}$ of the weight of 1 lb. in London, or about $\frac{1}{2}$ oz. weight, and the dyne is about $\frac{1}{981}$ of the weight of a gram.

The weight of a mass m is mg.

§ 7. Motion under uniform force.

In Chap. I, § 12, it was proved that

$$v - v_0 = at. \qquad (1)$$

$$v^2 - v_0^2 = 2as. \qquad (2)$$

Now if the moving mass is m, the acceleration a is due to a force F, such that

$$F = ma. \qquad (3)$$

If the mass m is known, the formulae (1)—(5) of Chap. I, § 12, suffice to find F; or if F be known, m can be found.

Writing $a = \frac{F}{m}$ in (1) and (2), we have

$$m(v - v_0) = Ft,$$

$$\tfrac{1}{2}mv^2 - \tfrac{1}{2}mv_0^2 = Fs.$$

When a mass m moves with velocity v, the product $\frac{1}{2}mv^2$ is called its kinetic energy.

If several forces all directed along the same straight line act on a body, the Second Law of Motion asserts that the acceleration is the sum of the accelerations which each force acting by itself would produce.

If a body of mass m be raised with acceleration a by some agent, the resultant force is equal to ma.

Let F be the upward force exerted by the agent, W the weight of the body, then

$$F - W = ma.$$

F is greater or less than W according as the upward velocity is increasing or diminishing.

When the motion can be supposed to take place with indefinite slowness the acceleration can be made as small as we please; the difference between F and W will be correspondingly small.

If therefore we are considering the transfer of a body from one point to a point on a different level, and are at liberty to indefinitely diminish the speed at which we suppose the transfer to take place, we can regard the force exerted as equal and opposite to the weight of the body.

Again if the motion takes place with uniform velocity

$$a = 0, \text{ and } F = W.$$

§ 8. Parallelogram of Forces.

We shall now consider the case when two or more forces act on a particle in different directions.

The Second Law of Motion, as stated in § 3, may be amplified as follows:—

The momentum generated by a force in a given time is proportional to the force and is in the same direction; and when several forces act on a particle, the momentum generated in a given time is obtained by compounding the momenta which each force acting separately would in the same time impart to the body.

Let a force P, acting by itself on a particle, generate a momentum OA in a given time, and let a force Q, acting by itself on the particle, generate a momentum OB in the same time.

Thus $$\frac{P}{OA} = \frac{Q}{OB}.$$

When both P and Q act together it follows, from the Second Law, that the momentum generated is represented by OC, the diagonal of the parallelogram $OACB$.

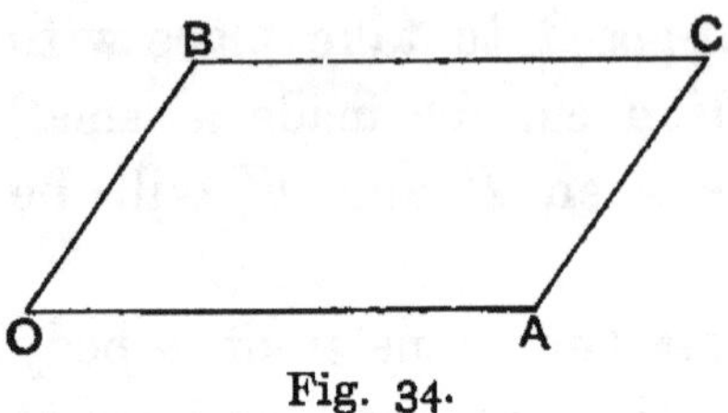

Fig. 34.

Now this momentum would be generated in the same time by a force R parallel to OC, such that

$$\frac{P}{OA} = \frac{Q}{OB} = \frac{R}{OC}.$$

And since R communicates the same momentum to the particle as P and Q do, it may be substituted for them and is called their resultant.

The resultant of two forces is therefore found by compounding them according to the Parallelogram Law.

Conversely, any force OC can be replaced by its components OA, AC or OA, OB, acting on the same particle.

Triangle of Forces.

A force represented by CO produces a momentum equal and opposite to that produced by OA and OB acting together.

Therefore the three forces CO, OA, OB acting together produce no momentum.

But CO, OA, OB are equal and parallel to the sides of the triangle COA taken in order.

Therefore if three forces, represented in magnitude and direction by the sides of a triangle taken in order, act on a particle, the particle will remain at rest; or, if already moving, it will continue in motion along a straight line with uniform velocity.

This proposition is called the Triangle of Forces.

Polygon of Forces.

Let forces acting on a particle be represented in magnitude and direction by the straight lines OA, AB, BC, CD not necessarily all in the same plane.

Then OB is the resultant of OA and AB; OC is the resultant of OB and BC, that is, of OA, AB, and BC; OD is the resultant of OC and CD, that is of OA, AB, BC and CD.

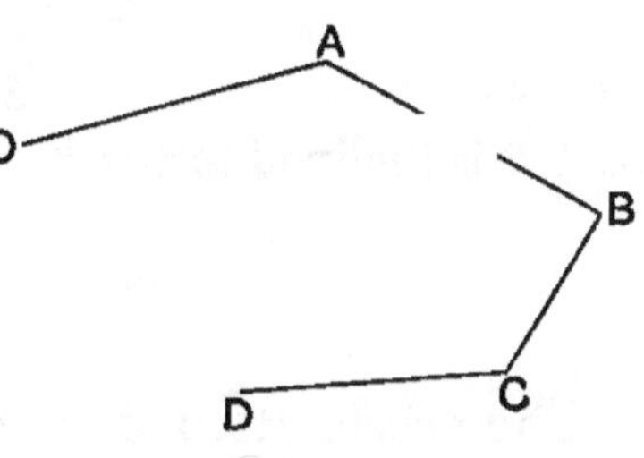

Fig. 35.

Hence if any number of forces acting on a particle are given in magnitude and direction, their resultant is obtained in the same way as the resultant of displacements represented by the same vectors as the forces.

Therefore, corresponding to the Polygon of displacements, we have the Polygon of Forces, viz.:—

If the forces which act on a particle are represented in magnitude and direction by the sides of a closed polygon taken in order, the particle remains at rest or in a state of uniform motion.

When several forces, all in one plane, act on a particle, their resultant is most easily obtained by resolving the forces along two perpendicular lines, in the manner explained previously for displacements.

Example.—To find the resultant of the following forces, 150 dynes N., 200 dynes N.E., 100 dynes E.

The most convenient directions in which to resolve are N. and E.

Since $\cos 45^\circ = \frac{1}{\sqrt{2}}$, the force of 200 dynes resolves into $100\sqrt{2}$ dynes N. and $100\sqrt{2}$ dynes E.

Therefore the total force *N.* is

$$150 + 100\sqrt{2} \text{ or } 50(3 + 2\sqrt{2}).$$

And the total force *E.* is

$$100 + 100\sqrt{2} \text{ or } 100(1 + \sqrt{2}).$$

And if R be the resultant force

$$R^2 = 2500\,(29 + 20\sqrt{2})\,;$$

whence $R = 378$ approximately.

And R is inclined to the E. at an angle a, such that

$$\tan a = \frac{3 + 2\sqrt{2}}{2 + 2\sqrt{2}} = \frac{1 + \sqrt{2}}{2}.$$

The acceleration due to the forces can also be found by this method, for let Ox, Oy be the two straight lines at right angles, m the mass of the particle, X the sum of the component forces parallel to Ox, Y the sum of the components parallel to Oy.

Then if a and b are the components of acceleration parallel to Ox and Oy, we have by the Second Law

$$ma = X$$

$$mb = Y.$$

Hence a and b can be found when m, X and Y are known.

Motion on a smooth Inclined Plane.

A smooth surface is one whose pressure on any body in contact with it is perpendicular to the surface.

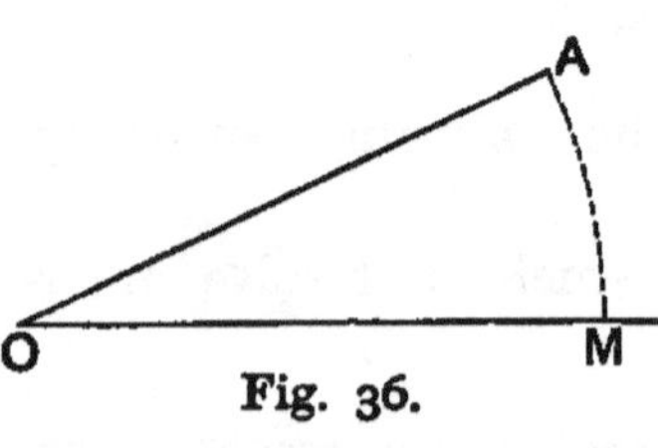

Fig. 36.

Let OA represent a line of greatest slope on the plane, OM a horizontal line in the vertical plane through OA, m the mass of a particle moving on the plane, a the angle AOM.

Resolve the weight of the particle along and perpendicular to the plane.

The latter component is $mg \cos a$ and balances the pressure of the plane.

The former component is $mg \sin a$, and is the only force in the plane of motion.

Therefore the particle has an acceleration $g \sin a$ down the line of greatest slope, and the formulae of Chap. I, § 13, apply, $g \sin a$ being substituted for g.

Thus the path is a parabola, unless the particle is projected along the line OA.

If v be the component of the initial velocity along OA, the height to which the particle will rise is $\dfrac{v^2}{2g \sin a}$.

The velocity perpendicular to the line of greatest slope, i.e. parallel to the horizontal line on the inclined plane, is constant.

§ 9. **Law III.** *To every action there is an equal and opposite reaction.*

Whatever pushes or pulls another body is pushed or pulled by it to the same extent. If pressure be applied to a piece of stone or clay with the finger, the stone or clay exerts an equal and opposite pressure on the finger; and if a horse draws a stone at the end of a rope, the stone exerts on the horse, by means of the rope, a force equal and opposite to that which the horse exerts to pull the stone along. If the sun exerts a force on the earth, the earth exerts an equal and opposite force on the sun.

The Law asserts that all force is the result of the *mutual* action between two portions of matter; in considering force applied to a single body, we only view one aspect of the phenomena due to it.

The mutual action of two portions of matter is called a Stress.

The Law holds whether the bodies between which there

is stress are at rest or in motion, and it also holds when one body is being deformed by the other, as when the clay is indented by the pressure of the finger.

Since equal and opposite forces generate equal and opposite momenta, it follows that, if two particles exert any action on each other, the total momentum of the two particles is not altered by this action. The same holds for any number of particles.

In the instance given above, the reader may possibly not see that our ideas are consistent with the fact that the horse moves forward, though the stone drags him back.

The explanation is that the horse's feet exert an action on the ground, which is partly downward, and partly backward, and the ground exerts an equal and opposite action on the horse, partly upward and partly forward, and the latter component, if it is greater than the tension in the rope, urges the horse forward.

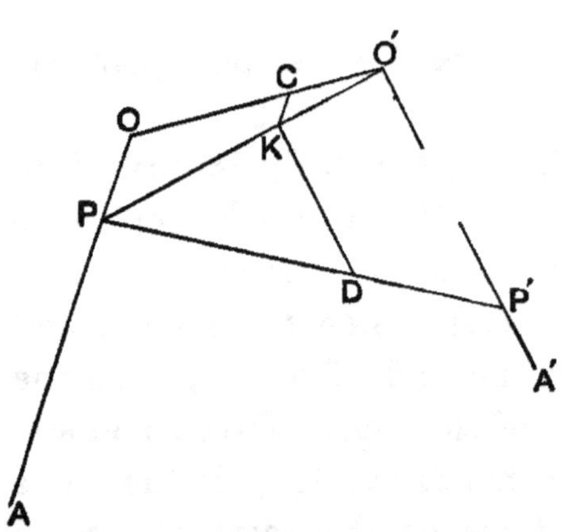

Fig. 37.

Centre of Mass.

Let two particles whose masses are M, M' move with velocities v, v' respectively along the lines OA, $O'A'$, which are not necessarily in the same plane.

Let O, O' be the positions through which the particles are passing at a given instant, P, P' the positions they would attain at the end of one second, if they continued moving with unchanged velocity. Then $OP = v$, $O'P' = v'$.

Let C be a point dividing the distance OO', so that

$OC : O'C :: M' : M$. C is then called the centre of mass of the particles at O, O'.

Join PP', $O'P$, and draw CK parallel to OP meeting $O'P$ in K, and KD parallel to $O'P'$ meeting PP' in D.

$$\text{Then } \frac{PD}{P'D} = \frac{PK}{O'K} = \frac{OC}{O'C} = \frac{M'}{M}.$$

Therefore D is the position which the centre of mass reaches in one second, and the velocity of the centre of mass is obtained by compounding CK, KD.

$$\text{Now } \frac{CK}{OP} = \frac{O'C}{OO'} = \frac{M}{M+M'};$$

$$\text{therefore} \quad CK = \frac{Mv}{M+M'}.$$

$$\text{Similarly} \quad \frac{KD}{O'P'} = \frac{PD}{PP'} = \frac{M'}{M+M'};$$

$$\text{therefore } KD = \frac{M'v'}{M+M'}.$$

The velocity of the centre of mass is therefore found by compounding the momenta of the two particles, and dividing the resultant momentum by $M+M'$.

Now suppose that the particles M, M' are acted on by any forces and also exert force on one another. In a given time they acquire momentum, but the momenta due to their mutual actions are equal and opposite.

Therefore the velocity acquired by the centre of mass is parallel and proportional to the resultant momentum due to external forces, i.e. to forces other than the mutual actions between the particles.

If there are no external forces, the centre of mass remains at rest or in motion with uniform velocity.

It is not necessary that the mutual action of the particles should be in the straight line joining them.

Example. A ball is fired horizontally from a cannon which is free to move on a smooth horizontal plane.

The only external forces acting on the system formed by the ball, cannon and powder, are their own weights and the resistance of the plane, which are all vertical, and do not affect the horizontal motion.

The combustion of the powder brings into play internal forces in the system, and generates a total horizontal momentum which must be zero.

Neglecting the mass of the powder, let M, m be the masses of the cannon and the ball, v the initial velocity of the ball.

Then V, the velocity of recoil, must be such that

$$MV + mv = 0,$$

$$\text{or} \quad V = -\frac{mv}{M}.$$

If the velocities of two particles, of mass M, m, are denoted by Oa, Ob respectively, the velocity of their centre of mass is denoted by Oc, where c is determined by the condition that $M \cdot ca = m \cdot bc$.

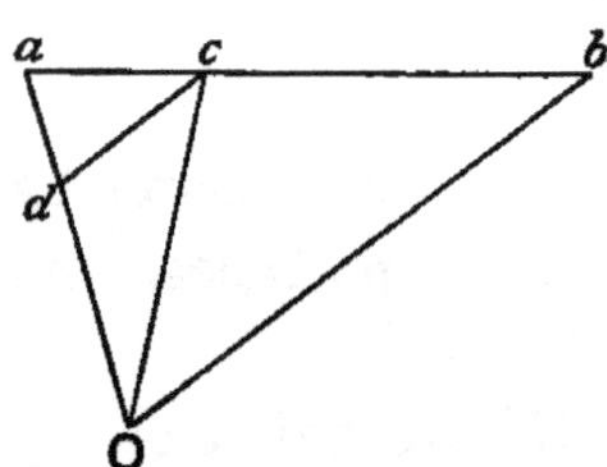

Fig. 38.

For if cd is parallel to Ob, Oc is compounded of Od and dc.

$$\text{And} \quad \frac{Od}{Oa} = \frac{bc}{ba} = \frac{bc}{bc+ca} = \frac{M}{M+m}.$$

Therefore
$$\frac{da}{Oa} = 1 - \frac{Od}{Oa} = \frac{m}{M+m};$$

and
$$\frac{dc}{Ob} = \frac{da}{Oa} = \frac{m}{M+m}.$$

Therefore Oc is the resultant of $\frac{M}{M+m} \cdot Oa$ and $\frac{m}{M+m} \cdot Ob$; that is, it is the velocity of the centre of mass.

§ **11. Centripetal force.**

A point moving with uniform velocity v in a circle of radius r has acceleration $\frac{v^2}{r}$ towards the centre. If the point be replaced by a particle of mass m, a force $\frac{mv^2}{r}$ towards the centre must be exerted on the particle to maintain its motion. This force is called Centripetal force, and must be supplied by some constraint; the circle may be a metal hoop which prevents the particle from continuing its motion in a straight line; or the particle may be fastened by a string to the centre of the circle; or may be acted on by a gravitating mass at the centre of the circle, as in the case of a planet moving round the sun. But whatever be the source of constraint, there must be a reaction equal and opposite to the action in question. The pressure of the particle on the hoop or its pull on the string is equal and opposite to the constraint exerted on the particle.

When a railway train passes round a curve, the rails exert on the train a pressure directed towards the centre of the curve, and the train exerts an equal and opposite pressure on the rails, in addition to the ordinary pressure which exists on a straight road. The pressure is proportional to the square of the velocity of the train, and inversely proportional to the radius of the curve; if the rails cannot stand so great a pressure, they give way and the train runs off the line. Thus to round a sharp curve at high speed is dangerous, and a heavy engine is more likely than a light one to run off the line on a curve, supposing both engines to have the same number of wheels.

Problems in uniform circular motion may often be simplified by introducing the conception of centrifugal force, which we shall now explain.

In a given position of the particle on the circle, let M, N be the components of the acting force taken along the tangent and radius of the circle. Then $N = \frac{mv^2}{r}$, $M = 0$.

But these can be written $N - \frac{mv^2}{r} = 0$, $M = 0$, which show that the particle would be in equilibrium under forces N, M and $\frac{-mv^2}{r}$; i.e. under the forces specified in the question together with a force $\frac{mv^2}{r}$, which we call centrifugal, since its direction is away from the centre.

Thus for the problem—To determine forces to which is due a motion with uniform velocity v in a circle of radius r; we can substitute the problem—To determine forces which together with a centrifugal force $\frac{mv^2}{r}$ maintain a particle in equilibrium at a given point of a circle of radius r.

Example. The upper end of a light string of length l is attached at O to a vertical rod which can turn freely about its axis OC, and a particle of mass m is attached to the lower end of the string. To find the angular velocity ω of the vertical rod in order that the string may make an angle α with the vertical.

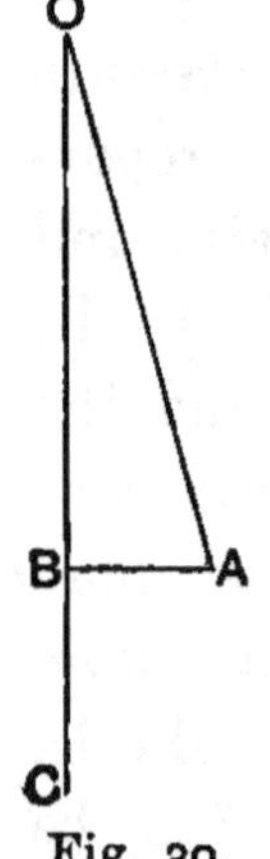

Fig. 39.

Let l be the length of the string OA, and draw AB perpendicular to OC.

The acting forces are T the tension of the string, and the weight mg.

Therefore the mass m would be maintained in equilibrium by the forces T, mg, and $m\omega^2 . BA$ respectively parallel to AO, OB, BA.

Therefore by the triangle of forces,

$$\frac{T}{AO} = \frac{mg}{OB} = m\omega^2.$$

But $$OB = l \cos \alpha.$$

Therefore $$\omega^2 = \frac{g}{l\cos a}.$$

The number of revolutions per second is $\frac{\omega}{2\pi}$ or $\frac{1}{2\pi}\sqrt{\frac{g}{l\cos a}}$.

The string and particle form a ***conical pendulum.***

When $\omega < \sqrt{\frac{g}{l}}$ the string does not quit the vertical.

The above result also holds with fair accuracy when instead of a string we have a light rod hinged horizontally at its upper end to the vertical rod, and instead of the particle a heavy ball.

Watt's Governor.

This is employed to check accidental variations in the speed of an engine; it consists of two conical pendulums hinged horizontally on a vertical cylinder.

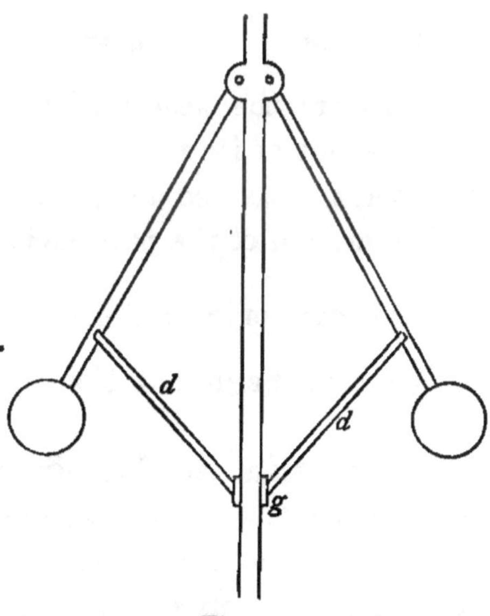

Fig. 40.

The cylinder is kept in rotation about its axis by the engine, and its angular velocity bears a constant ratio to the speed that is to be controlled, this relation being maintained by a train of mechanism.

When the speed increases the balls rise, and as they do so the ring g connected with them through the rods dd is raised, and sets in motion mechanism by which the supply of steam is decreased; when, on the contrary, the speed diminishes, the balls fall in and displace g in the opposite direction, increasing the supply of steam.

§ 12. Time of Oscillation of a Simple Pendulum.

Let P be a heavy particle suspended from a fixed point O by a string OP, the mass of which is negligible in comparison with that of the particle. Draw OM vertical, and PM perpendicular to OM.

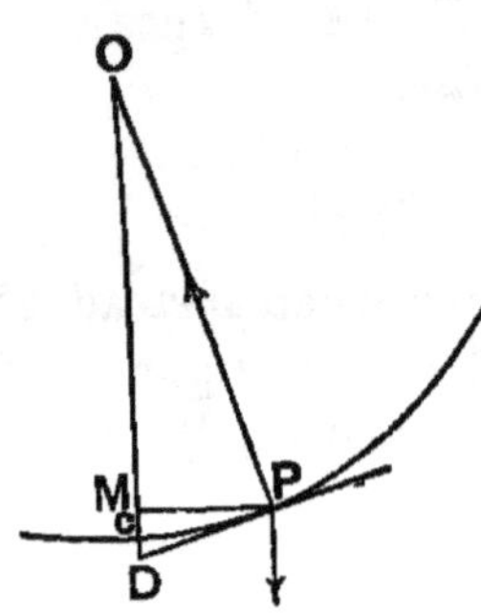

Fig. 41.

Let $OP = l$, $POM = \theta$.

Then if a is the component of the acceleration of P along PC, we have, by the Second Law of Motion,

$$a = g \sin\theta = g\,\frac{PM}{l}.$$

But when the arc CP is made very small,

$$PM = \text{arc } PC.$$

Therefore $$a = \frac{g}{l} \text{ arc } PC.$$

Therefore the acceleration of the particle towards the lowest point of its path is proportional to its distance from this point, measured along the arc.

The motion of the particle is consequently (Chap. I, § 17) a simple harmonic motion of period $2\pi\sqrt{\frac{l}{g}}$.

We have neglected the effect of the resistance of the air.

To ascertain whether different substances fall to the earth with the same acceleration, Newton took several equal small boxes and suspended each by a thread eleven feet long, and placed in each box different substances as metal, wood, &c. He thus had a series of simple pendulums whose motions were equally resisted by the air.

Failing, after careful observation, to find any difference in the times of oscillation, he inferred that g was the same for all bodies.

As the oscillations can be compared with great accuracy by observing the time of a large number, this is much the most satisfactory way of proving that in vacuo all bodies fall to the earth with the same acceleration.

We shall now determine a limit to the error committed by regarding an oscillation through an angle α from the vertical as an indefinitely small oscillation.

Let the tangent at P meet OMC at D.

$$\text{Then} \quad a = g\frac{PM}{l} = g\frac{PD\cos\theta}{l}.$$

$$\text{And} \quad PM < \text{arc } PC < PD.$$

$$\text{Therefore} \quad a < \frac{g}{l}\text{ arc } PC, \quad \text{and} \quad > \frac{g\cos\theta}{l}\text{ arc } PC;$$

$$\text{or} \quad < \frac{g}{l}\text{ arc } PC, \quad \text{and} \quad > \frac{g\cos\alpha}{l}\text{ arc } PC;$$

for $\cos\alpha < \cos\theta$, except at the extremity of the oscillation, when $\cos\theta = \cos\alpha$.

Therefore the time of vibration lies between the periodic times of two harmonic vibrations—

$$a = \frac{g}{l}\text{ arc } PC, \quad \text{and} \quad a = \frac{g\cos\alpha}{l}\text{ arc } PC.$$

Therefore the time of vibration of a pendulum which oscillates through an angle α from the vertical lies between

$$2\pi\sqrt{\frac{l}{g}} \quad \text{and} \quad 2\pi\sqrt{\frac{l}{g\cos\alpha}}.$$

We have already seen that $\cos\alpha$ approaches very nearly to 1 when α is moderately small (Chap. I, p. 45).

A more accurate value of the time of vibration through an angle α can be found by using the Calculus.

§ 13. Atwood's Machine.

The essential part of Atwood's Machine consists of a pulley mounted so as to revolve freely on an horizontal axis, and a light string passing over the pulley.

From the ends of the string hang two equal pans in which masses can be placed.

In order to diminish the effect of friction on the motion of the pulley, each end of the axle is mounted on two friction wheels. These revolve freely in fixed bearings mounted on a stand, and as their motion is slow compared with that of the axle, the resistance from friction is insignificant, and the weights of the suspended masses are the only external forces which need be considered.

The stand which carries the friction wheels is mounted on the top of a pillar about seven feet high, to which a vertical scale graduated in inches or centimetres is attached.

We shall suppose that the string is inextensible, and that the tension in it is the same throughout. It will appear afterwards that the latter assumption is not quite justifiable.

Let m, m' be the suspended masses, m being the greater, and let T be the tension in the string, a the acceleration of m downwards. Since the string is inextensible, m' has the same acceleration upwards.

By the Second Law of Motion,

$$ma = mg - T, \quad m'a = T - m'g.$$

Therefore $$a = \frac{m - m'}{m + m'} g,$$

$$T = \frac{2\,mm'g}{m + m'}.$$

Thus the greater mass descends with increasing velocity, and the acceleration may be diminished by increasing the masses in each pan, keeping their difference about the same.

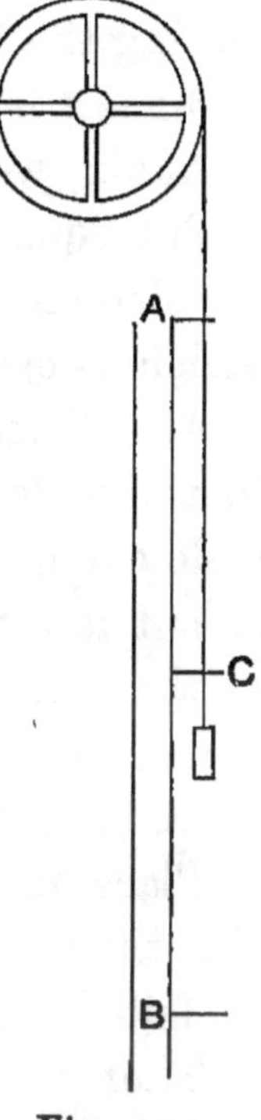

Fig. 42.

Verification of the Laws of uniformly Accelerated Motion, by Atwood's Machine.

At the zero of the vertical scale is placed a stop A, the removal of which permits the mass m to fall.

Another stop B which arrests the fall can be placed at any other point, and the distance fallen is measured by the reading on the scale at B.

The time of fall can be determined by a stop-watch or water-clock.

In order to verify the formula $s = \frac{1}{2} at^2$, we suspend masses m and m' from the strings, and observe the times of fall from A to different positions of B.

Then, if s be any one of these distances and t the corresponding time; $\frac{s}{t^2} = \frac{1}{2} a$.

Now a is the same throughout. Therefore $\frac{s}{t^2}$ should be the same for all values of AB.

Since $$a = \frac{m - m'}{m + m'} g,$$

g can be found by this experiment, but the method is not satisfactory, even when a correction to be noticed in Chap. IV is introduced. Resistances due to friction are appreciable, and it is best to find g by timing the oscillations of a given pendulum (see p. 71).

We may also vary the value of a by transferring masses from one pan to the other, and verify the formula by observing the times required to fall through a given distance.

Since s is constant, we have $at^2 =$ constant.

Again, we can verify the formula $v^2 = 2\,as$.

Put equal masses in the pans, and place on the top of the descending pan a small bar of mass μ, the ends of which project beyond the sides of the pan.

At a distance from A equal to the chosen value of s attach to the scale a slider C carrying a horizontal ring, which is wide enough to allow the descending pan to pass through it, but is too small to admit the bar μ. As the pan passes C the bar is removed, the masses in the pans become equal, and the pans move with uniform velocity unless there is sensible resistance from friction.

Place the stop B a convenient distance below C, start the time-keeper when μ is removed, and note the time of fall from C to B.

Then the distance BC being known, v is known.

If the experiment is made for different values of AC, the acceleration is constant, and v^2 is proportional to s; thus, if AC has the values 1, 4, 9 feet, the corresponding values of v are as 1, 2, 3.

Examples.

1. A mass M rests on a smooth horizontal table; a string attached to M passes over a small pulley at the edge of the table and supports a mass m hanging freely. To determine the acceleration and the tension in the string.

Let a be the acceleration, T the tension.

Since the tension in the string is the only horizontal force acting on M,

$$T = Ma.$$

Also, since the acceleration of m is vertically downwards, and the vertical forces acting on m are its own weight mg, and the tension of the string, $$mg - T = ma.$$

Therefore $$a = \frac{mg}{m+M}, \quad T = \frac{Mmg}{m+M}.$$

2. Masses of 3, 4, and 5 lbs. respectively move along the sides of an equilateral triangle ABC taken in order, with velocities 25, 20, and 18 feet per second. To find the resultant momentum of the whole system and the force which would generate this momentum in 10 seconds.

The momenta are 75 along AB, 80 along BC, 90 along CA; these reduce to $75 - 80$ along AB, and $90 - 80$ along CA, or to 5 along BA, and 10 along CA.

10 along CA can be resolved into 5 along BA and $5\sqrt{3}$ perpendicular to BA.

Therefore the resultant momentum is $\sqrt{(5+5)^2+(5\sqrt{3})^2}$ or $\sqrt{175}$ or $5\sqrt{7}$,

and the required force is $\frac{5\sqrt{7}}{10}$ or $\frac{1}{2}\sqrt{7}$ poundals.

3. A mass m describes the perimeter of an equilateral triangle ABC with a velocity x of uniform magnitude. Find the impulse of the blow which m receives at each vertex of the triangle.

Let the particle pass at B from AB to BC.

Its momentum is mx, initially along AB, finally along BC.

The change of momentum is obtained by reversing the initial momentum, and then compounding with the final.

Now the resultant of mx along BA, and mx along BC, is $mx\sqrt{3}$ along the internal bisector of the angle CBA.

Therefore the impulse of the blow is $mx\sqrt{3}$, and it bisects the angle CBA.

4. A ball of mass 10 lbs. projected upwards with a velocity of 30 feet per second, strikes a ceiling 9 feet above the point of projection and penetrates 4 inches into it. Determine its momentum at the time of striking, the energy of the blow, and the mean resistance to penetration. ($g = 32$ foot-second units.)

The velocity v at the time of striking is given by
$$v^2 = 900 - 2 \times 32 \times 9 = 324, \text{ or } v = 18.$$

The momentum is therefore 180.

And the energy of the blow is $\frac{1}{2} \cdot 10\, v^2 = 1620$.

The mean resistance to penetration being F, and s the distance 4 inches, we have (§ 7),

$$Fs = 1620, \quad \text{and} \quad s = \tfrac{1}{3};$$

$$\therefore \quad F = 4860 \text{ poundals, or } 151\tfrac{7}{8} \text{ pounds' weight.}$$

5. A particle of mass 6 oz. is attached to one end of an inextensible string, which passes over a smooth pulley and has attached to its other end a mass of 3 oz. To the latter mass is attached a second inextensible string which hangs downwards and terminates in a mass of 2 oz.

Neglecting the masses of the strings, to find the acceleration of the system and the tensions in the strings.

Let T_1, T_2 be the tensions, a the acceleration.

The first string has a mass of 6 oz. hanging from one end, and 3 + 2 or 5 oz. hanging from the other.

Therefore the motion of the mass of 6 oz. or $\frac{3}{8}$ lb., gives

$$\tfrac{3}{8}a = \tfrac{3}{8}g - T_1,$$

and the motion of the mass of 5 oz. gives

$$\tfrac{5}{16}a = T_1 - \tfrac{5}{16}g.$$

$$\therefore \quad a = \frac{6-5}{6+5}g = \tfrac{32}{11} \text{ foot-second units,}$$

$$\text{and} \quad T_1 = \tfrac{3}{8}g\,(1 - \tfrac{1}{11}) = \tfrac{15}{44}g \text{ poundals.}$$

To find T_2.

The mass of 2 oz. hanging from the second string moves upwards with acceleration $\frac{1}{11}g$.

$$\therefore \quad T_2 = \tfrac{1}{8}(1 + \tfrac{1}{11})g = \tfrac{3}{22}g \text{ poundals.}$$

6. A train whose mass is 112 tons is travelling at a rate of 25 miles per hour on a level track, and the resistance due to air, friction, &c. is 16 lbs. weight per ton. A carriage of mass 12 tons becomes detached. Assuming that the force exerted by the engine is the same throughout, find how much the train will have gained on the detached part after 50 seconds, and the velocity of the train when the detached part comes to rest, given that

$$g = 32.$$

After the carriage is detached, the resultant force on the train is 16×12 lbs. weight, and the force on the carriage is equal and opposite to this.

Therefore the acceleration of the train is $\frac{16 \times 12 \times 32}{100 \times 2240}$ or $\frac{96}{3500}$ foot-second units; similarly the acceleration of the carriage is $-\frac{32}{140}$.

Therefore the acceleration of the train relative to the carriage is $\frac{32}{125}$, and the distance gained is

$$\frac{1}{2} \times (50)^2 \times \frac{32}{125} = 320 \text{ feet.}$$

Let x be the speed of the train in miles per hour when the detached carriage comes to rest.

The momentum is then $100 \times 2240 \times x$.

The momentum before parting is $112 \times 2240 \times 25$, and these momenta are equal, since the forces exerted on the train and detached carriage are equal and opposite. Therefore

$$100x = 112 \times 25,$$

or $x = 28$ miles per hour.

7. The time of fall from rest down all chords of a vertical circle to the lowest point is the same.

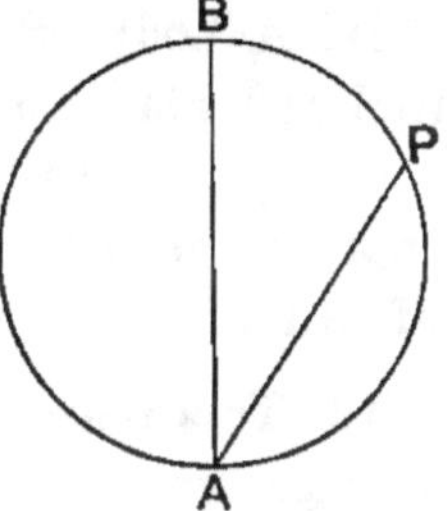

Fig. 43.

Let AB be a vertical diameter, P any point on the circle, a the radius of the circle, $PAB = \theta$.

The acceleration of a heavy particle falling along AP is $g \cos \theta$.

Also $AP = 2a \cos \theta$.

Hence (Chap. I, § 12), if t be the time of fall from P to A,

$$t = \sqrt{\frac{4a \cos \theta}{g \cos \theta}} = 2\sqrt{\frac{a}{g}}.$$

Thus the time of fall is the same for all positions of P on the circle APB.

8. A mass of 10 lbs. is whirled round 5 times in a second on a smooth horizontal table at the end of a fine string of length 5 feet. Calculate the tension in the string.

The velocity is $2\pi \times 5 \times 5 = 50\pi$.

The radius of the circle is 5.

Therefore the acceleration towards the centreis

$$\frac{2500\pi^2}{5} \text{ or } 500\pi^2,$$

and the tension in the string is $500\pi^2 \times 10 = 5000\pi^2$ poundals.

9. A body flung vertically upwards with a velocity of 2695 cm. per sec. is stopped by a board after moving for 4 seconds. What is its kinetic energy just before stopping, and with what impulse does it strike the board, its mass being $3\frac{1}{2}$ grams? $(g = 980.)$

10. A body passes a point at height 54·5 cm. with a velocity of 436 cm. per sec.: with what initial velocity was it thrown up, and how long will it rise? Also, what is its kinetic energy at the given point, its mass being 49? $(g = 981.)$

11. From a point O within a triangle ABC, straight lines OD, OE, OF are drawn perpendicular to the sides of the triangle and proportional to their lengths. Show that forces represented by OD, OE, OF are in equilibrium.

12. A body whose mass is 25 lbs. is supported on a smooth inclined plane, of length 50 feet and height 14 feet, by a force P acting parallel to the plane. Determine the value of P, and show that the pressure of the body on the plane is equal to the weight of 24 lbs.

13. To a mass of 507 grams is fastened one end of each of two strings, whose other ends are fastened to two nails in the same horizontal line. The lengths of the strings are 10 and 24 decimetres, and they form a right angle with each other. Find (1) the tension of each string, (2) the portion of the weight which each nail supports, (3) the force with which each nail is drawn towards the other.

14. A cord, having a mass of 10 kilos. attached to each end, passes over two smooth pullies placed at A and B in the same horizontal line, 16 centimetres apart, and also through a very small ring C from which a mass of 12 kilos. is suspended. Find the depth below AB at which the ring C will rest.

15. Forces $5P$, $13P$ act at a point, at such an angle that the direction of the resultant is perpendicular to that of the smaller force. Find the magnitude of the resultant.

16. $ABCD$ is a square, AC its diagonal. Forces of magnitudes 3, $3\sqrt{2}$, 5 act at A in the directions AB, AC, AD respectively: find the magnitude of their resultant, and show that its line of action meets CD in a point E such that $3\,CE = ED$.

17. *ABC* is an isosceles triangle right-angled at *C*; particles whose masses are 10, 15, and 20 grams are respectively moving along the sides *AB*, *BC*, *CA* with velocities 15, 10 and 5. Find the resultant of the momenta of the particles.

18. A 6 oz. ball strikes a bat with velocity 10 feet per sec., and returns with velocity 30 feet per sec.; if the duration of the impact be $\frac{1}{20}$ second, find the average force exerted by the striker.

19. A mass of 6 lbs. hangs by two strings, one making an angle 60°, the other 30°, with the vertical. Find the tension in each string.

20. A body weighing 3 lbs. is projected vertically upwards with a velocity of 40 feet per sec.; find the numerical values of its velocity, acceleration, kinetic energy, and momentum at the end of 1 sec. ($g = 32$ foot-second unit.)

21. A mass of $2P$ lbs. is fastened to a weightless string, which passes over a pulley, and begins to lift a mass of P lbs. After the first mass has fallen freely for 1 sec., how high will the second mass be lifted?

22. A force of 2 poundals acts on a mass of 30 lbs. for two minutes. What momentum, and what kinetic energy, will be generated?

23. A mass of 10 lbs. on a smooth horizontal table is connected by a weightless string passing over a smooth pulley at the edge of the table with a mass of 5 lbs. which hangs vertically. Find the acceleration of either mass, and the tension of the string in lbs. weight.

Also if the mass of 10 lbs. be 3 feet from the edge of the table, find how long it takes to pull it over.

24. How many oscillations does a pendulum of length 31·5 cm. make in 242 seconds, where $g = 981$ C. G. S. units?

25. A simple pendulum performs 7 complete vibrations in 9 seconds; or shortening it by 28.35 cm., it performs 7 complete vibrations in 5 seconds. Hence determine g. ($\pi = 3\frac{1}{7}$.)

26. A body of mass 10 lbs. moves with uniform velocity in a circle whose radius is 5 feet. The only force acting upon it is

equal to 25 lbs. weight. Find the velocity with which the body moves. ($g = 32$ foot-second units.)

27. A smooth circular tube of radius a in a vertical plane contains a heavy particle, and is set in uniform rotation with angular velocity ω about a vertical axis through its centre. Find the position of the particle when it is at rest relatively to the tube.

28. A string of length 2 feet has its upper end fastened to a fixed point O and to the lower end a particle of mass 4 lbs. is attached, which revolves with uniform velocity about the vertical through O. Supposing the string just able to bear the weight of 8 lbs., find the greatest inclination of the string to the vertical.

29. A mass of 2 lbs. is suspended from a fixed point by a string 5 feet long. How must it be projected so as to describe a horizontal circle whose plane is 1 foot below the fixed point? Find the tension in the string.

30. A horizontal board moves up and down with a simple harmonic motion, the time of a complete oscillation being 1 sec. Find the greatest amplitude admissible that a weight placed on the board may not be jerked off.

31. Two bodies of mass 40 oz. and 50 oz. are suspended from the ends of a string which passes over the pulley of Atwood's machine. Through what distance will motion take place in the first three seconds, and what will be the velocity of the moving masses at the end of that time?

32. A light inelastic string passes over a light pulley, and has masses of 9 oz. and 12 oz. attached to its ends. On the 9 oz. mass a bar whose mass is 7 oz. is placed, which is removed by a fixed ring after it has descended 7 ft. from rest. How much further will the 9 oz. mass descend? ($g = 32$ foot-second-units.)

33. Three precisely similar nails stand partly driven into a board of uniform hardness, the exposed portion of each being of length 1 inch. One of them is now hit with a hammer-head of $\frac{1}{2}$ lb. mass moving with a velocity of 10 feet per second, and is then just completely driven in. Find (supposing the resistance of the

board to penetration to be independent of the portion of a nail which has already penetrated) :—

(1) From what height a hard mass of 1 lb. must be dropped on the head of the second nail to drive it in completely;

(2) What mass placed at rest on the head of the third nail is the least that will make it penetrate?

CHAPTER III.

WORK AND ENERGY.

§ 1. Work done by a Force.

When a particle is displaced through a distance s under the action of a force F in the direction of displacement, the force is said to do Work on the particle, which is measured by the product Fs. This statement holds whether other forces concur in causing the displacement or not.

If the force and displacement are in opposite directions, as when a heavy body moves upwards against the earth's attraction, they must be considered as of opposite sign, so that the work done by the force is negative.

It is sometimes convenient to consider work as done against a force. In estimating work done *against* a force, the force and displacement have the same sign when they are in *opposite* directions.

Work is appropriately measured by the product Fs. For it is naturally regarded as proportional to F, and the exertion of a force through a distance $2s$ may be looked on as consisting of two successive exertions through a distance s. A force does no work when the particle on which it acts remains stationary.

The unit of work in the C. G. S. system of units is the **Erg.** It is the work done by a dyne in displacing the point at which it acts through a centimetre.

In the English system two units have been used: the **Foot-Poundal**, which is the work done by a poundal in moving the point at which it acts through a foot; and the **Foot-Pound**, which is the work done against gravity in raising a pound through the height of a foot.

The foot-pound is g foot-poundals. Since its magnitude depends on the latitude and height above sea-level, it is not suitable for a scientific unit.

Since the foot-pound-second unit of velocity is 30·48 times the C. G. S. unit, and the pound is 453·6 grams, the poundal is $453{\cdot}6 \times 30{\cdot}48$ dynes and the foot-poundal is $453{\cdot}6 \times (30{\cdot}48)^2$ or $4{\cdot}214 \times 10^5$ ergs.

The foot-pound is 32·19 foot-poundals at Greenwich, or $1{\cdot}3564 \times 10^7$ ergs.

Examples. 1. Express in ergs and in foot-pounds the work done against gravity in raising 105 grams through 20 centimetres.

2. Express similarly the work done in raising 25 lbs. through 10 feet.

3. The masses hung from the string of Atwood's machine being 9 oz. and 5 oz., calculate in foot-poundals the work done when the larger mass has fallen 5 feet.

Definition of Power.

The power of an agent is the rate at which it can do work, measured by the work done per unit time.

Engineers have used the Horse-Power (denoted by H. P.) as unit of power; this is the power of an agent which does 550 foot-pounds of work per second.

The modern electrical unit is the Watt, which is the power exerted in doing 10^7 ergs per second.

Since 1 H. P. $= 550 \times 1{\cdot}3564 \times 10^7$ ergs per second,

1 H. P. $= 746$ Watts approximately.

When the displacement and the force are not along the same straight line, the displacement is resolved into its components along and perpendicular to the line of action of the force, and the product of the former component by the force is taken as the work done. Thus, a force does no work when the displacement is perpendicular to it.

Let OF and Of represent the force and the displacement, and draw fe, FE perpendicular to OF and Of respectively.

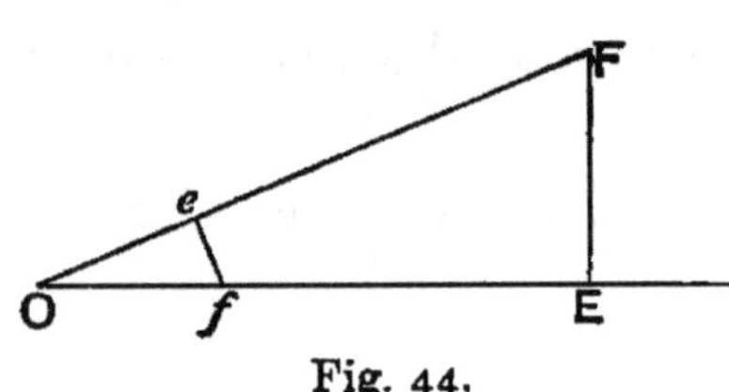

Fig. 44.

Then a circle can be described about $FEfe$ (Euclid III, 24), and therefore the rectangles $OF \,.\, Oe$, $OE \,.\, Of$ are equal.

Hence the work done by a force can also be represented by the product of the displacement and the component of the force in the direction of displacement.

If several forces act on a body, the total work done is estimated by the sum of the quantities of work done by each force.

§ 2. The work done by two forces in any displacement is equal to the work done by their resultant.

Let the forces which act on a particle be represented in magnitude and direction by the sides of a closed polygon taken in order. The sum of the components of these forces in any direction, and therefore in the direction of displacement, is zero. Therefore forces which maintain a particle in equilibrium do no work in any displacement.

Let the forces L, M, N, acting at O, be represented by the sides PQ, QR, RP of a triangle PQR.

Then, in any displacement, the work done against PQ is equal to that done by QR and RP.

But the work done against PQ is equal to the work done by QP in the same displacement.

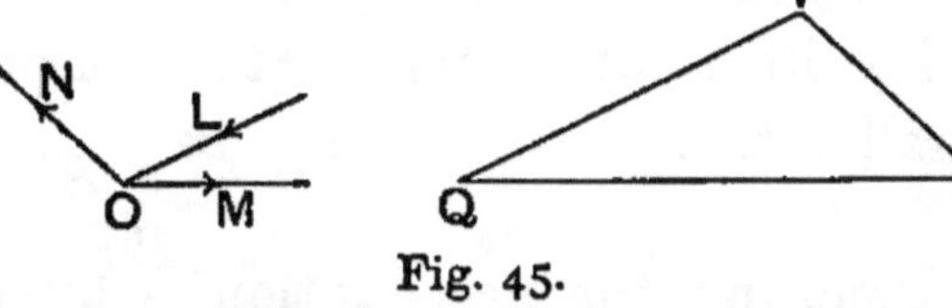

Fig. 45.

Therefore the work done by QP is equal to that done by QR and RP. But QP represents the resultant of QR and RP. Therefore the work done in any displacement by two forces acting at a point is equal to the work done by their resultant.

The same result is clearly also true for any number of forces greater than two.

§ 3. In any motion of a particle the work done by the acting force is equal to the gain of kinetic energy.

(*a*) Rectilinear motion under uniform force.

It has been shown that if F be the force acting on a mass m; v_0, v the velocities at the beginning and end of the time during which m moves through a distance s,

$$Fs = \tfrac{1}{2} m (v^2 - v_0^2).$$

Now $\frac{1}{2} mv_0^2$, $\frac{1}{2} mv^2$ are the initial and final values of the kinetic energy of m. Thus, in rectilinear motion under uniform force, the work done is equal to the gain of kinetic energy.

(*b*) Curvilinear motion under uniform force. This case includes the motion of a projectile under gravity.

Let v_0 be the initial velocity of the moving particle m,

P_0 and p_0 its components parallel and perpendicular to the direction of the force F,

$$v_0^2 = P_0^2 + p_0^2.$$

Let v be the velocity acquired in any displacement, P and p its components, s the component of displacement parallel to F,

$$v^2 = P^2 + p^2.$$

Denoting the acceleration $\frac{F}{m}$ by a, we have (Chap. I. § 13)

$$2\,as = (P^2 - P_0^2),$$

$$p = p_0.$$

Therefore

$$Fs = \tfrac{1}{2}m\,(P^2 - P_0^2) = \tfrac{1}{2}m\,(P^2 + p^2 - P_0^2 - p_0^2) = \tfrac{1}{2}m(v^2 - v^2_0).$$

Here also the work done is equal to the gain of kinetic energy.

(*c*) Motion under a variable force.

The diagram on which in Chap. I we expressed the distance traversed in a given time with variable velocity, also enables us to express the work done by a variable force.

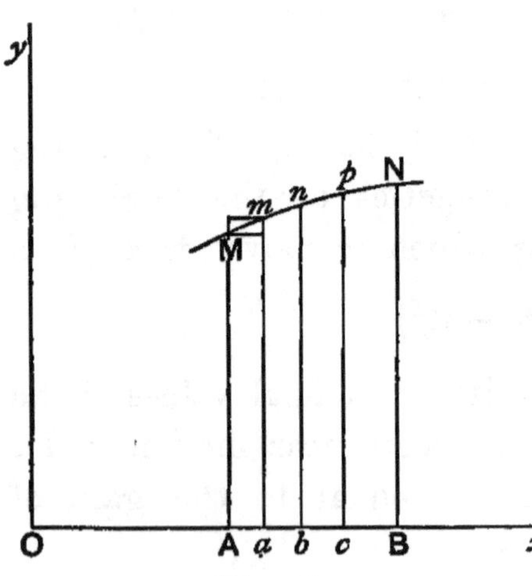

Fig. 46.

Let Ox, Oy be two perpendicular lines, and let the displacement of the particle be represented by a distance measured along Ox.

From A, representing any position of the particle, draw the ordinate AM denoting the force in the direction of motion which acts on the particle at A.

For all other positions of the particle let similar ordinates be drawn; their extremities will lie on the curve $MmnpN$.

As the particle passes from A to a, the working force varies from AM to am, and the work done lies between $Aa \, . \, AM$ and $Aa \, . \, am$, i.e. between the areas aM and Am. Similar limits can be found for the work done in the successive displacements ab, bc, cB. Hence the work done in the whole displacement lies between two magnitudes, one of which is slightly greater, and the other slightly less, than the area $AMNBA$.

But, by increasing the number of segments between A and B, it can be proved, as in Chap. I, § 9, that the work done and the area $AMNBA$ both lie between two quantities which can be made as nearly equal as we please, and therefore the work done and the area $AMNBA$ are themselves equal.

In the displacement Aa, the force increases from AM to am; and the kinetic energy generated lies between that due to a uniform force AM, and that due to a uniform force am; it is therefore represented by an area intermediate between Am and aM. Similarly the kinetic energy generated in a displacement ab lies between am and bn, and similar results can be found for the energy generated in later displacements.

Therefore the total kinetic energy, generated by a variable force in a displacement from A to B, lies between the same limits as the work done, and it has been shown that the difference $Aa\,(BN - AM)$ between the limits can be indefinitely diminished.

Hence the work done by this force is equal to the kinetic energy generated.

§ **4. Central Forces.**

Let us consider the motion of a particle under a force which is always directed towards a fixed point, called the centre of force. The particle does not move along a straight line, unless it is initially either at rest or in motion along the line joining it to the centre.

In the most important cases in nature, the force depends only on the distance of the particle from the centre of force; and two cases are particularly important, namely those in which the force is proportional, (1) to the distance of the particle from the centre, (2) to the inverse square of the distance from the centre.

Let C be the centre of force, μ the force exerted on the particle when it is at unit distance from C. The force on the particle when it is at distance x from C is, in the first case μx, in the second case $\frac{\mu}{x^2}$, directed towards C.

Motion under a force proportional to the distance from C, the centre of force.

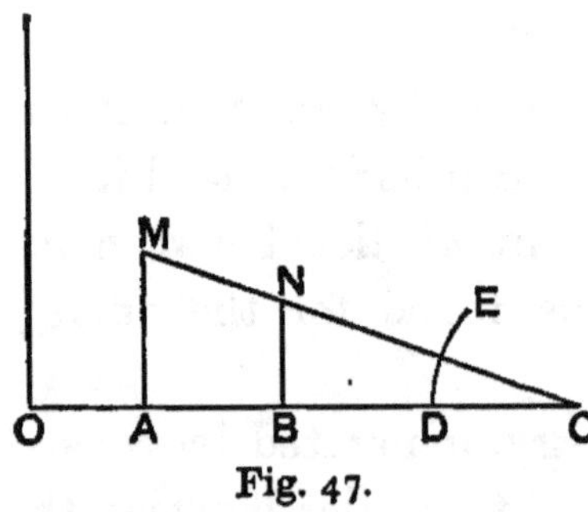

Fig. 47.

If m is the mass of the particle its acceleration is $\frac{\mu x}{m}$ towards C. Therefore when it starts from rest at A its motion is a simple harmonic motion along CA, and the curve $MmnpN$ [§ 3, (*c*)] becomes a straight line MN passing through C, for $AM = \mu CA$ and $BN = \mu CB$.

The area $AMNB$

$$= \tfrac{1}{2}(AM + BN)AB = \frac{\mu}{2}(CA + CB)(CA - CB)$$

$$= \frac{\mu}{2}(CA^2 - CB^2).$$

If v be the velocity of the particle when it passes B, $\frac{1}{2}mv^2$ is its kinetic energy, and this is equal to the work done.

Therefore $v^2 = \frac{\mu}{m}(CA^2 - CB^2)$.

The same expression for v^2 also follows from Chap. I, § 17.

If V be the velocity at a point D on CB,

$$V^2 = \frac{\mu}{m}(CA^2 - CD^2).$$

And $$V^2 - v^2 = \frac{\mu}{m}(CB^2 - CD^2).$$

Motion under a central force proportional to the inverse square of the distance from C.

We shall prove in Chap. VI, that if $\frac{\mu}{x^2}$ is the attraction to C of a particle at a distance x from C, the work done in the displacement of the particle from A to B is

$$\mu\left(\frac{1}{BC} - \frac{1}{AC}\right).$$

§ 5. Let a particle be compelled to describe its path by some frictionless constraint, e. g. a smooth tube in which the particle moves.

The force due to the constraint is then always perpendicular to the direction of motion of the particle, and does no work. If there are no other forces acting on the particle, its velocity is uniform; and if other forces exist, they do work equal in amount to the gain of kinetic energy.

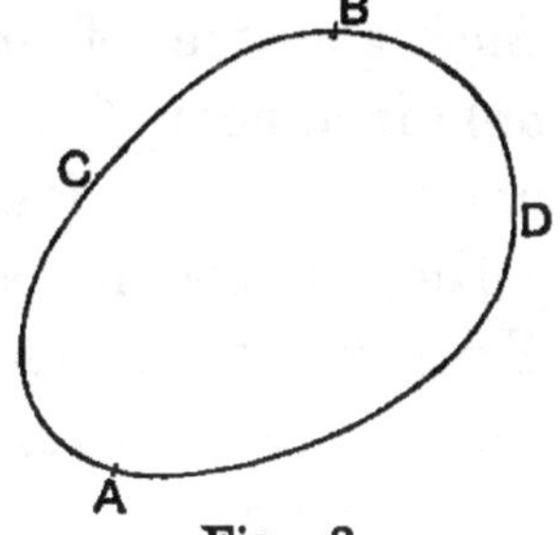

Fig. 48.

Now if A is the initial and B the final position of the particle, the displacement from A to B may be caused in an infinite variety of ways, simply by varying the constraint to which the particle is subject. For example, if the particle move in a tube, we may give the channel of the tube any form we please.

Let ACB, ADB be any two of these paths forming a circuit $ACBDA$, and let the external forces be such that the kinetic energy gained in passing from A to B by any path is the same as that lost in returning from B to A along the same path.

Then the kinetic energy gained in passing from A to B must generally be the same for all paths along which the passage can be made. For otherwise if the kinetic energy gained in passing from A to B along ADB exceeds that gained along ACB, the particle will, after describing the circuit $ADBCA$, possess greater kinetic energy than it did at starting, and can augment this energy indefinitely by repeatedly describing the circuit.

Now, with a remarkable exception in electricity, natural forces do not exhibit this phenomenon, and therefore forces which are not altered when the direction of motion changes are generally such that the work they do and the kinetic energy they generate depend only on the initial and final positions of the particle and not on the path by which the particle makes its journey.

Such a system of forces is called a Conservative system; simple instances of it are the weight of a body, the mutual attraction of the Planets and Sun, the attractions and repulsions of electric charges and of magnetic poles.

Frictional forces (so called) are generally non-conservative, for they change their direction when that of the motion changes.

We may now conveniently sum up the results obtained in the preceding sections as follows:

If a particle undergoes a displacement from A to B, under the action of known forces which may be constant or variable, the path described by the particle can be varied to any extent by varying the magnitude and direction of the initial velocity at A. But whatever this velocity may be, the gain of kinetic energy in passing from A to B is equal to the work done by the acting forces; and, if the acting forces belong to a conservative system, the gain of kinetic energy is the same *for all paths* between A and B.

The path of the particle between A and B may also be varied by the introduction of constraints which do no work, e.g. by causing the particle to move along a smooth tube.

§ 6. If at all points of the closed path the velocity is the same, the kinetic energy is constant, and the external force is either zero or is perpendicular to the direction of motion. If the kinetic energy is not constant, let it be least when the particle is at A and greatest when it is at B.

On any curve between A and B take a point P, such that the work done on the particle in its passage from A to P is c. By taking all the curves which can possibly be drawn from A to B, we find an infinite number of positions of P, for we have at least one for every curve from A to B. These points all lie on a surface which is called a level surface.

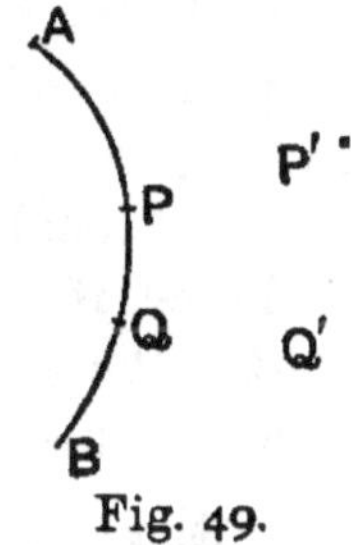

Fig. 49.

Let Q be a point in AB such that d is the work done in passing from A to Q. Then, taking all the curves that can be drawn from A to B, we find that all the positions of Q lie on another level surface.

Let P' be any point on the first level surface, Q' a point on the second. The work done in the displacement $P'Q'$ is $d-c$, for the work done in passing from P' to A is $-c$, and from A to Q' is d.

This vanishes if $d = c$. Hence in passing from one point to another of the same level surface no work is done.

Two level surfaces cannot cut one another, for if they did the work done in passing from A to one of their common points would have two different values, which is impossible.

When a particle moves freely under its own weight, the level surfaces are horizontal planes, if the earth's curvature can be neglected.

Illustrations.

(1) If a particle of mass m is projected from a point P in any direction with any velocity, and crosses at Q a horizontal plane at a depth h below P, the work done by gravity in the fall of the particle is mgh, and this is equal to the gain of kinetic energy. This statement holds good for any position of Q on the horizontal plane, for any initial velocity, and for any constraints under which the particle moves, provided that the force exerted by the constraint is perpendicular to the direction of motion.

(2) If the bob of a pendulum is raised to a height h above O, its position of equilibrium, and is then allowed to fall, the increase of kinetic energy in passing to O is the same as if the bob fell freely through a height h.

Hence, if v is the velocity at O, $v^2 = 2\,gh$.

At O the tension in the string and weight of the bob are both perpendicular to the direction of motion. Hence the acceleration is vertical, and is equal to $\frac{v^2}{l}$, or $\frac{2\,gh}{l}$, where l is the length of the pendulum.

(3) Consider again the case of motion under a force proportional to the distance from the centre (§ 4).

We have proved that when the motion takes place along CB (Fig. 47) the work done in a displacement from B to D is

$$\tfrac{1}{2}\mu\,(CB^2 - CD^2).$$

If E is any other point such that $CE = CD$, D and E are on the same level surface, and no work is done in passing from D to E.

Therefore the work done in a displacement from B to E by any path is $\frac{1}{2}\mu(CB^2 - CD^2)$ or $\frac{1}{2}\mu(CB^2 - CE^2)$, and this is also the kinetic energy acquired.

Work done by an Impulse.

The work done in increasing the velocity of a mass m from v_0 to v is $\frac{1}{2} m(v - v_0)(v + v_0)$, and $m(v - v_0)$ is the impulse of the acting force if the motion is rectilinear. Hence in this case the work done by an impulse I in changing the velocity of a particle from v_0 to v is

$$\tfrac{1}{2} I(v_0 + v).$$

Motion of two or more particles.

Since as before stated when a single particle is displaced by a force, its gain of kinetic energy is equal to the work done on it, let us now consider the displacement of two particles by the action of force, and see if we are led to the same conclusion.

Let A, B be the initial positions of the particles P and Q, a, b their final positions; and let there be a stress T between the particles tending to separate them. We shall suppose that the stress either is constant, or depends only on the distance between the particles.

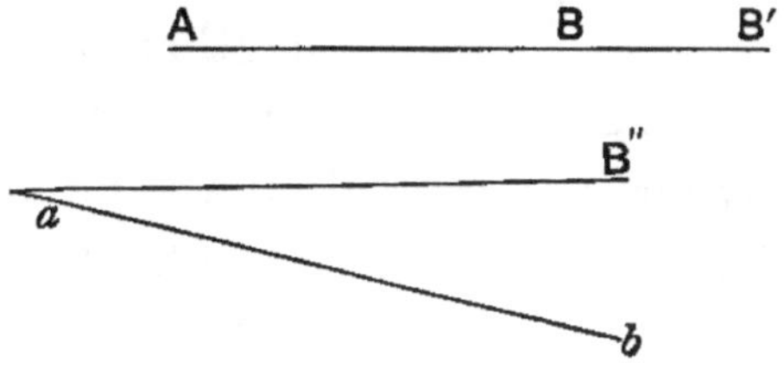

Fig. 50.

We may consider the displacement as made by the following operations:

(1) P remaining fixed at A, Q is displaced along AB to a point B' such that $AB' = ab$.

(2) P is displaced to a, and Q is displaced from B' to B'', where aB'' is equal and parallel to AB'.

(3) P remains fixed at a, and Q is brought from B'' to b by a motion of rotation round a.

No work is done on the whole by T in the second and third displacements, for in (2) T does as much work on P as is done against T on Q, and in (3) P remains fixed and Q moves at right angles to T.

We have then only to consider the first displacement.

If T is constant the work done is $T.BB'$.

If T is variable the work is more difficult to express, but when T depends only on the distance AB, it is a conservative force, and does the same amount of work in producing the given displacement, whatever be the operations by which we suppose the displacement effected.

The stress T still existing, let X be the resultant of other forces acting on P, Y the resultant of other forces acting on Q. X and Y are called external forces.

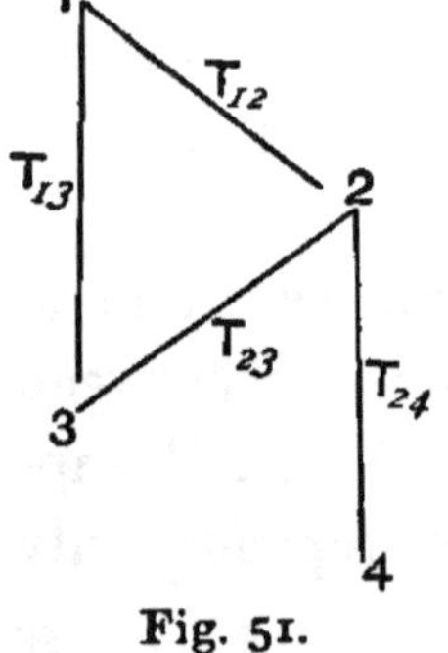

Fig. 51.

The kinetic energy gained by P and Q in any displacement is equal to the work done on P and Q by the forces X and Y and the stress T, and therefore the work done by the forces X and Y is equal to the kinetic energy gained by P and Q, together with the work done against T.

Consider for instance two balls, connected by an elastic string, and pulled apart by forces applied to the balls. Here the stress T tends to bring the balls together. If after displacement the balls are at rest, the work done by the external forces is precisely equal to the work done against the tension of the string in stretching it. But if the balls are still moving, the work done by the external forces is equal to the kinetic energy of the balls and the work done in stretching the string.

Similarly, if there be any larger number of particles, 1, 2, 3, 4, ..., moving under external forces and under their mutual actions T_{12}, T_{23}, T_{13}, ..., the total work done by all the forces is equal to the kinetic energy imparted to the system, and can be divided into work done by the external forces and by the mutual actions. Therefore the work done by the external forces is equal to the kinetic energy gained, together with the work done against internal forces.

A material system considered with respect to the relative positions of its parts is said to have a configuration, and a system of particles is said to have the same configuration when the relative positions of the several particles are the same. The configuration is said to be changed when the relative positions of the particles are altered.

Potential Energy.

If the particles of a material system are displaced from their position of equilibrium by the action of external forces, the work done by the external forces is expended partly in increasing the kinetic energy, partly in doing work against the internal forces.

Now let the external forces cease to act. The constrained condition which they brought about—as in stretching the string in the example above—is no longer possible, and the particles tend to return to the configuration from which they started. Let the system pass (if necessary under the guidance of constraints which do no work) to its original configuration. Then if the internal forces form a conservative system, they do work (say W) equal to the work that was spent against them, and increase the kinetic energy by W.

Thus, when work is done against the internal forces of a system, the system acquires a power of generating kinetic energy—by reverting to its former configuration—which it did not possess before. The existence of this power is expressed by the statement that the system has Potential Energy.

The potential energy of a system in any configuration is the work that its internal forces (which are supposed to form a conservative system) can do in bringing it from this to a standard configuration; and the loss (or gain) of potential energy, in any displacement, is measured by the work that is done by (or against) the internal forces in that displacement.

Hence if the particles of a system move under their mutual actions only, the sum of the kinetic and potential energies of the system remains constant throughout the motion. If external forces act on the system, the work which they do is equal to the increase of the total energy, kinetic and potential, of the system. If the particles are so connected together that in any displacement their configuration is unchanged, no work is done by the internal forces.

If a system consists of a single particle, its energy is entirely kinetic.

An extended rigid body is a collection of particles whose configuration does not change. Therefore if external forces are applied to a rigid body the work done by them is equal to the kinetic energy communicated to the body.

As an illustration of potential energy, let us consider a wire stretched by the application of force to its ends, and let W be the work spent in stretching the wire. Had the same work been spent in moving the wire as a rigid body, it would have acquired kinetic energy W.

Now let the stretching force be removed, the wire contracts rapidly and vibrates, its energy becoming partly kinetic and partly potential; the total energy is still W.

If the internal forces were conservative and the air did not resist the motion, the wire would vibrate for an indefinitely long time. As these conditions are not realised, it speedily comes to rest.

§ 7. Conservation of Energy.

The Principle of the Conservation of Energy is as follows:—

The total energy of a system of particles cannot be increased or diminished by the mutual actions of the parts of the system.

The total energy in the material universe is not susceptible of either increase or decrease.

An increase or decrease in the energy of a system of particles is due to the performance of an equivalent amount of work by or against external forces.

Example. Motion of a falling body relatively to the earth.

Let m be the mass of the body, W its weight, M the mass of the earth.

For simplicity let us suppose that the earth is initially at rest and does not rotate on its axis.

Since action and reaction are equal and opposite, m exerts a force W on the earth. We suppose that the earth is subject to no other forces, and we regard the earth and the body as forming a material system.

Let V, v be the velocities upward acquired by the earth and body, in approaching nearer by a distance h.

Then, by the Third Law of Motion,

$$MV + mv = 0.$$

And by the Conservation of Energy $\frac{1}{2}MV^2 + \frac{1}{2}mv^2 = Wh.$

Substituting for V its value $-\frac{mv}{M}$, we have

$$v^2 = \frac{2Wh}{m\left(1 + \frac{m}{M}\right)}.$$

Since $\frac{m}{M}$ is a fraction which is quite insensible, the velocity V is insensible; and the centre of mass of the earth and body, which remains fixed, practically coincides with the centre of the earth.

We may therefore generally regard the earth as fixed when considering questions relative to falling bodies, and we may take the kinetic energy acquired by a falling body as equal to the work done by its weight in the fall.

§ 8. Let us consider the behaviour of the simple pendulum as deduced from the principle which we have laid down.

If the string is inextensible the energy of the bob at any moment is wholly kinetic, and is due to the work done by gravity on the bob in its descent. This energy is a maximum when the bob is moving fastest, i.e. when it is passing its position of equilibrium; but immediately after this position is passed the bob begins to rise, does work against gravity, and loses all its energy, coming to rest for an instant at the height from which it began to fall. The bob then falls back, acquires the same kinetic energy as before in reaching its lowest position, and then loses it again in rising against gravity to the position from which it started originally. We have now traced the motion through a complete oscillation, and all subsequent oscillations resemble this.

Thus, according to theory, the pendulum continues for an indefinite time to oscillate though a constant angle from the vertical.

In practice we know that this is not so. The times of successive oscillations are equal, but their amplitudes steadily diminish.

The explanation of this gives us the opportunity of considering the extension of the principle of conservation of energy to the case when non-conservative forces act.

§ 9. Our statement of the Principle of the Conservation of Energy has only been established for the case when the internal forces form a conservative system. Work which has been spent against non-conservative forces is not recovered as kinetic energy when the motion is reversed.

Thus, in the rise of the pendulum bob from its lowest position, work is done against the friction of the air as well as against the force of gravity; on reversing the motion the latter part of the work is recovered as kinetic energy, but not the former. The kinetic energy of the pendulum is thus wasted, and the pendulum ultimately comes to rest.

It may appear then that the Principle only holds for a certain class of motions, and that one which is not very common in our experience. The Principle, however, has a much wider significance, and as this is a very important conclusion, we shall briefly state the evidence for it.

About the end of the last century considerable doubt was thrown on the then generally accepted theory that heat was an impalpable fluid pervading material bodies. The term 'fluid' expressed only the general belief that an increase of heat in any system could only arise from the disappearance of an equal amount of heat elsewhere.

This belief is consistent with some of the simple phenomena of calorimetry observed in the determination of specific heat by the method of mixtures, where it is true (under certain conditions) that no heat disappears from the system.

The belief is more difficult to reconcile with the evolution or absorption of heat on solidification or fusion, and other phenomena prove indisputably that heat can be generated where none existed before, and that existing heat can be made to disappear.

Boyle showed that a piece of iron can by hammering be heated till to touch it is intolerable; the heat thus generated can be communicated to other bodies, and the iron made to return to its original condition. Rumford in 1798 showed that in the boring of cannon enough heat can be generated in $2\frac{1}{3}$ hours to boil 19 lbs. of water in addition to the heating of the casting and machinery.

Davy rubbed two pieces of ice together in a receiver, taking every precaution that they might not receive heat from external sources. He found that some of the ice was melted. Now the water resulting from the ice certainly contains more heat than it did when solid, and the unmelted ice is unaltered. Thus heat has appeared which cannot be traced to previously existing heat.

In 1843 the late Dr. Joule began his researches on the Mechanical Equivalent of Heat. He began by showing experimentally that if a copper cylinder with an iron core is rotated between the poles of an electromagnet, more heat is generated in the cylinder and more work is required to drive it when the battery circuit is made than when it is broken. The appearance of this heat cannot be accounted for by the disappearance of heat elsewhere.

Different experiments showed that the additional heat

generated measured in calories* bore a constant ratio to the additional work done measured in ergs or foot-pounds.

In a second set of experiments, Joule measured the heat evolved by the sudden compression of air in a closed receiver under the action of a known pressure. The work done in the compression was also determined, and was found to bear the same relation to the heat evolved, as that obtained before.

The most elaborate experimental proof of the law followed in 1849. A paddle driven by falling weights rotated on a vertical axis in a vessel of water. Projecting vanes were fixed inside the vessel which just allowed the paddle to pass, but, by impeding the flow of water, greatly increased the difficulty of driving the paddle.

The amount of work required could be estimated from a knowledge of the weights and the distance through which they fell.

The temperature of the water was found to be raised by the motion of the paddle, and the number of calories of heat imparted to the water was in different experiments strictly proportional to the work expended.

The numerical ratio obtained in this experiment agreed fairly well with the other results, and is entitled to the most weight since this apparatus is the most sensitive.

Experiments made with sperm-oil and mercury instead of water gave the same result.

The results of these experiments and of a fourth based on the measurement of the heat developed by an electric current, convince us that by the expenditure of work an equivalent amount of heat can be generated.

The amount of work required to generate a calorie is

* The calorie is the quantity of heat required to raise the temperature of unit mass of water by 1°.

called the Mechanical Equivalent of Heat. To raise the temperature of 1 lb. of water by 1° F. about 772·5 foot-pounds must be expended.

Illustrations of the Conservation of Energy.

Many phenomena lead us to believe that the molecules of all bodies are in a state of rapid motion, and that the effect of communicating heat to a body is to increase the energy of this motion.

Heat is communicated to us from the sun through apparently empty space, and it is found to take the same time as light to pass from one place to another. As this energy disappears from the sun and reappears on the earth after a time proportional to the distance travelled, we conclude that energy travels with uniform velocity from the sun to us, and exists in some form between us and the sun. As we have not yet succeeded in forming a conception of energy apart from matter, we hold that space is pervaded by a highly elastic medium of very great tenuity, called the ether, and that energy travels from the sun to us by means of the vibrations of this medium. Energy in the form of ether-vibrations is called Radiation, or Radiant Energy.

A sounding body, such as an organ pipe, excites vibrations in the air which travel in all directions from the body; it requires a supply of energy to keep it sounding, for the vibrations of the air which cause sound are a form of energy.

One of the most conspicuous sources of heat is combustion; when carbon and oxygen unite there is a great evolution of heat. Hence the tendency of two substances to unite chemically, or *chemical affinity*, may be regarded as potential energy.

This tendency is employed in electric batteries. If the

terminals of a cell are connected by a wire, a circuit is formed in which a current flows. The acid in the cell oxidises (or burns) the zinc, and potential energy is lost proportional to the amount of zinc dissolved. This energy is employed in maintaining the electric current.

Now when a current flows in a wire it may do work in several ways. It can exert force on other conductors conveying currents, or on magnets, and by displacing them it does work. It can also decompose, under proper conditions, the salts of metals, and as it gives the constituents the potential energy of their chemical affinity, it does work. It also generates heat in the wire which conveys it, and in the battery. By the conservation of energy the total energy obtained from the current in unit time is equal to that given by the combustion of the zinc in the battery.

Hence if we take two similar circuits maintained by batteries in which there is the same consumption of zinc per unit time, and if we allow the current in one circuit to decompose salts and to move magnets while the other current does no work, the latter current will generate more heat than the former.

The use of the dynamo depends on the fact that when a conducting circuit revolves in a certain manner among magnets or currents, a current is generated in the revolving circuit, which may be employed in the ways mentioned above. In accordance with the Principle of Conservation of Energy, more work is required to move the revolving circuit among magnets or currents than to move it when it revolves by itself, and the additional work employed is the equivalent of the energy obtained from the dynamo, appearing partly in the form of useful work as in lighting, electroplating, &c., and partly in the useless form of heat, noise, vibration, &c.

The reader will find an excellent general account of the Principle of the Conservation of Energy in two of Helmholtz's Popular Scientific Lectures, 'On the Conservation of Force,' and 'On the Interaction of Natural Forces.'

§ 10. Impact.

Consider the collision of two small spherical balls A and B moving along the line which joins their centres in the direction AB, the motion of each ball being one of translation only. Let m, m' be their masses, u, u' their velocities before impact, v, v' the velocities after impact; $u > u'$.

The total momentum is the same before and after impact.

Therefore $$mu + m'u' = mv + m'v'. \qquad (1)$$

As soon as the balls come into contact A exerts a pressure on B tending to drive it forward, and B exerts an equal and opposite pressure on A.

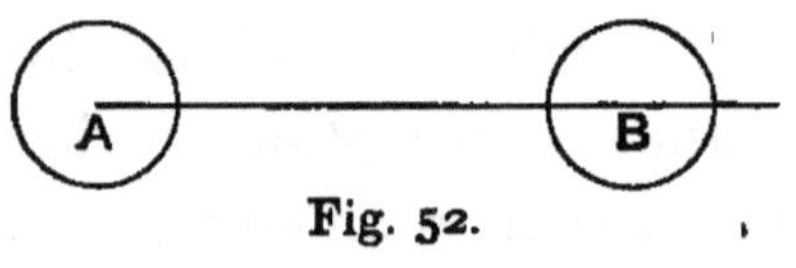

Fig. 52.

Now all bodies are really more or less deformed by the application of pressure.

As soon therefore as the balls come into contact with each other they begin to compress one another and continue to do so till the velocities of the balls are equal; at this instant the kinetic energy of the system is diminished by the work done in deforming the balls. But after this instant, the centres of the balls begin to separate again, and the mutual pressure does work on them and increases the kinetic energy.

Let us suppose that the kinetic energy finally returns to its initial value.

Then $$\tfrac{1}{2}mu^2 + \tfrac{1}{2}m'u'^2 = \tfrac{1}{2}mv^2 + \tfrac{1}{2}m'v'^2. \qquad (2)$$

And by (1) and (2), $$u - u' = v' - v. \qquad (3)$$

Whence the relative velocity of the balls after impact is the same as before in magnitude but is of the opposite sign.

If the ball B is at rest initially, $u' = 0$.

In this case $v = \frac{m - m'}{m + m'} \cdot u$.

Therefore if A impinges directly on a ball B at rest, A continues to move in the same direction if its mass is greater than that of B, while its motion is reversed if its mass is less than that of B. If the two balls are of equal mass, A is brought to rest and B moves on with velocity u.

The conditions implied above are never realised in nature, for the work done in the first part of the collision is expended in setting the balls in vibration (producing sound, heat, &c.) as well as in compressing the balls. Though the latter part of the energy may be recovered in the later part of the collision, the former is wasted and is converted directly or ultimately into heat.

Newton proved experimentally that the formula (3) may be replaced by

$$v' - v = e(u - u'), \tag{4}$$

where $e < 1$ for all substances.

e is called the Coefficient of Restitution; it is approximately the same for all velocities of the colliding balls, and depends only on the materials of the balls.

Combining (1) and (4) we have

$$(m + m')u = m'v'\left(1 + \frac{1}{e}\right) + v\left(m - \frac{m'}{e}\right),$$

$$(m + m')u' = mv\left(1 + \frac{1}{e}\right) + v'\left(m' - \frac{m}{e}\right).$$

Therefore

$$mu^2 + m'u'^2 = mv^2 + m'v'^2 + \frac{mm'}{m + m'}(e^{-2} - 1)(v - v')^2.$$

Therefore the kinetic energy lost in the collision is

$$\frac{mm'}{2(m+m')}(e^{-2}-1)(v-v')^2,$$

or $$\frac{mm'}{2(m+m')}(1-e^2)(u-u')^2.$$

The Impact is here said to be direct.

When $e = 1$, the balls are sometimes said to be perfectly elastic.

Oblique Impact of two smooth balls.

Let A and B be the positions of the centres of the balls m and m' at the moment of collision, and let the paths of the centres before collision lie in the same plane with AB and make angles a and a' with it.

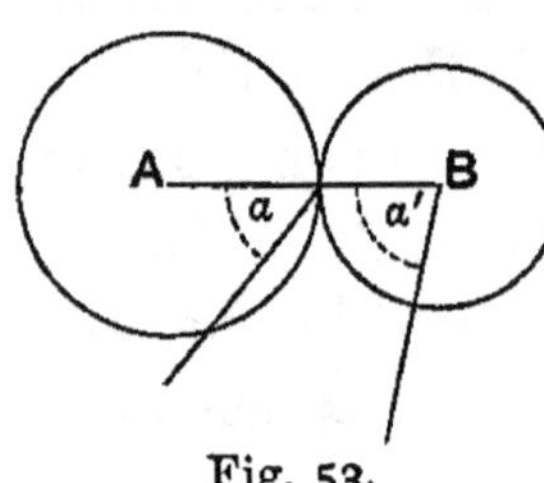

Fig. 53.

The impact is then said to be oblique.

Let u, u' be the initial velocities, v, v' the final velocities, β, β' the angles which the paths after collision make with AB.

Since the balls are smooth the stress between them during collision is along AB, and the components of velocity perpendicular to AB are not affected by the collision.

$$\text{Therefore} \quad u \sin a = v \sin \beta, \tag{5}$$

$$\text{and} \quad u' \sin a' = v' \sin \beta'. \tag{6}$$

Again, the total momentum along AB remains unaltered. Therefore

$$mu \cos a + m'u' \cos a' = mv \cos \beta + m'v' \cos \beta'. \tag{7}$$

Also the relative velocity of the balls along AB is affected in the same way as before.

$$\text{Therefore } v' \cos \beta' - v \cos \beta = e(u \cos a - u' \cos a'). \tag{8}$$

The equations (7) and (8) are of the same form as (1) and (4) with $u \cos a$, $u' \cos a'$,... written for u, u',...

Therefore

$$mu^2 \cos^2 a + m'u'^2 \cos^2 a' = mv^2 \cos^2 \beta + m'v'^2 \cos^2 \beta'$$
$$+ \frac{mm'}{m+m'}(1-e^2)(u \cos a - u' \cos a')^2.$$

Therefore by (5) and (6),

$$mu^2 + m'u'^2$$
$$= mv^2 + m'v'^2 + \frac{mm'}{m+m'}(1-e^2)(u \cos a - u' \cos a')^2.$$

And the kinetic energy lost is

$$\frac{mm'}{2(m+m')}(1-e^2)(u \cos a - u' \cos a')^2.$$

Impact on a fixed smooth plane.

Let a small ball moving with velocity u along BA impinge on a fixed plane at B.

Then if a is the angle between BA and BN, the perpendicular to the plane, the components of velocity along and perpendicular to NB are $u \cos a$, $u \sin a$. Let BC be the path of the particle after impact.

If x, y are the components of its velocity along and perpendicular to BN,

$$x = eu \cos a,$$
$$y = u \sin a.$$

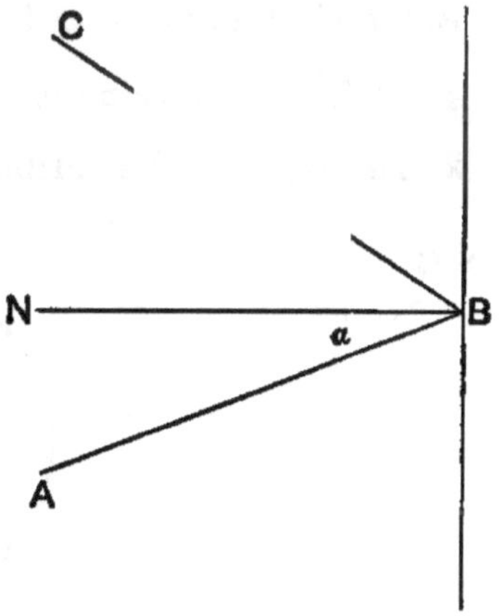

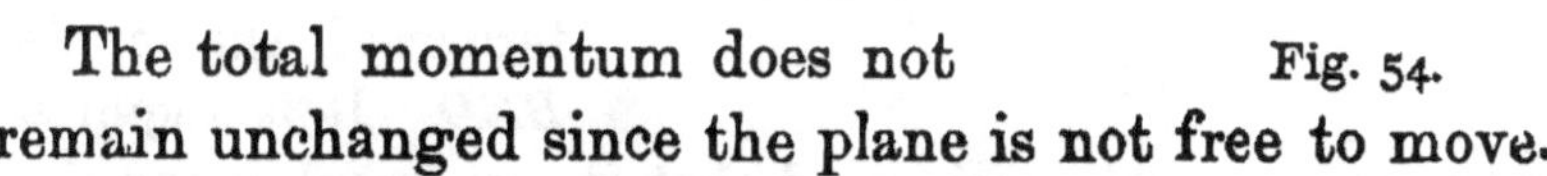

Fig. 54.

The total momentum does not remain unchanged since the plane is not free to move.

The resultant velocity after impact is

$$u\sqrt{\cos^2 a + e^2 \sin^2 a},$$

and $$\tan NBC = \frac{y}{x} = \frac{1}{e} \tan a.$$

If $e = 1$, the angles NBA, NBC are equal.

The impulse on the particle is given by its total change of momentum, and is $u \cos a \, (1 + e)$ parallel to BN.

Examples.

1. To find the H. P. of an engine which will raise 20 gallons of water per minute from a depth of $18\frac{1}{2}$ feet, and discharge it with a velocity of 100 feet per second.

The gallon of water weighs 10 lbs. and $g = 32$.

Therefore the work done per second in raising water is

$$\frac{20 \times 10 \times 32}{60} \times \frac{37}{2} = \frac{5}{3} \times 1184 \text{ foot-poundals.}$$

And the kinetic energy generated is

$$\frac{1}{2}\,\frac{20 \times 10}{60} \times (100)^2 = \frac{5}{3} \times 10000.$$

Therefore the whole work done is

$$\frac{5}{3}(1184 + 10000) = 18640 \text{ foot-poundals.}$$

But 1 H. P. $= 550 \times 32$ foot-poundals per sec.

Therefore the engine is $\dfrac{18640}{550 \times 32} = 1\frac{13}{220}$ H. P.

2. To explain the principle of the centrifugal railway, and to find the least height from which the carriage may start.

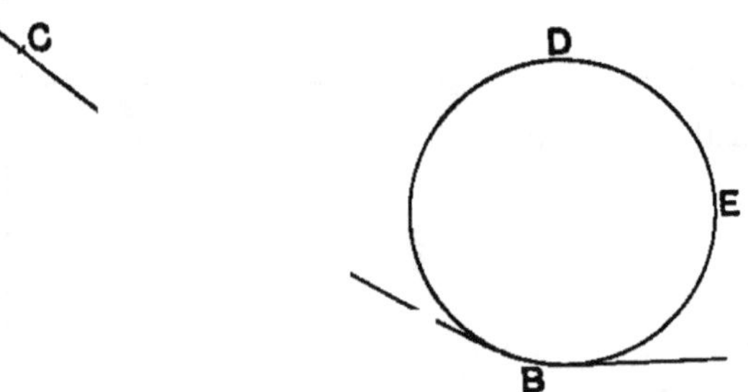

Fig. 55.

In the centrifugal railway, a carriage A runs down the inclined track CB and thence round the interior of the circle BED. It is required to start the carriage with such speed that it can travel round the circle without dropping off.

Let v be the velocity of the carriage at D, r the radius of the circle.

Then $\dfrac{v^2}{r}$ is the acceleration of the carriage, and is directed vertically downwards.

And the forces acting on the carriage are its own weight, and the resistance of the rail at D.

The acceleration due to these cannot be less than g.

Therefore $\frac{v^2}{r}$ must not be less than g, for otherwise the particle would fall from the circle.

Hence the velocity at D must be at least $\sqrt{rg}$.

At other points on the circle the velocity is greater than at D, and the radius is inclined to the vertical. Therefore if the carriage does not fall at D, it passes all other points safely.

Now assuming the railway to be smooth, the velocity at D is equal to that at C on the same level as D.

And the velocity acquired in falling through a height h is

$$\sqrt{2\,gh}.$$

Therefore h, the vertical distance through which the carriage descends from rest to C, cannot have a smaller value than that given by

$$\sqrt{rg} = \sqrt{2\,gh},$$

or

$$h = \tfrac{1}{2}r;$$

i. e. the carriage must descend from a point whose height above the highest point of the circle is at least $\frac{1}{2}r$.

3. A simple pendulum of length l is held horizontally and then let go. To find the velocity when the string is vertical and its tension T in this position.

The velocity v of the bob is given by $v^2 = 2\,gl$.

And at the lowest point the acceleration towards the centre is

$$\frac{v^2}{l} \quad \text{or} \quad 2\,g.$$

Therefore by the Second Law of Motion,

$$\frac{mv^2}{l} = T - mg, \quad \text{or} \quad T = 3\,mg.$$

4. A bullet weighing $\frac{1}{2}$ oz. moving with a velocity of 1200 feet per sec. strikes a bank of earth, and penetrates it to a depth of 4 feet; find the work done and the mean resistance to penetration.

5. A man of weight 160 pounds is in a swing whose ropes are 18 feet long; if he rises through an angle of 60°, what will

be the stress on the ropes when they are vertical? Give the answer in lbs. weight.

6. A smooth tube is in the form of the arc of a quadrant, of which one extreme radius is horizontal and the other turned vertically downwards. A thread passes up the tube and over its upper edge which is smooth. To the ends of the thread are fastened heavy particles one of which, m, is just within the lower end of the tube, while the other, p, hangs freely. Find the ratio of p to m, that p may draw m just up to the upper end of the tube.

7. How long must a 10 H. P. engine work to impress on a mass of 12000 tons initially at rest an ultimate velocity of 3300 yards an hour?

8. A smooth circular tube in a vertical plane contains a heavy particle. Find the least velocity with which the particle must be projected from the lowest point, in order that it may travel continually round the tube.

9. Find the H. P. of an engine which can start a train of 200 tons, and raise its speed to 30 miles an hour in $1\frac{1}{2}$ minutes.

10. A railway truck, the mass of which is 6400 lbs., stands on a horizontal railway, the total friction when it is in motion being $\frac{1}{64}$th of the weight. A man begins to push it with a force of 120 lbs. weight. Find the acceleration; find also the velocity and the rate of working after t seconds.

Show that if his highest rate is $\frac{1}{10}$ H. P., he cannot go on pushing with the initial force for more than about 4.6 secs.

11. A ball is projected with a velocity 120 feet per sec., at an elevation 60°, against a vertical smooth wall, 120 feet distant: where will it impinge? If the coefficient of elasticity be $\frac{1}{2}$, when and where will it strike the ground?

12. A 6 oz. ball moving with velocity 20 meets and strikes a 4 oz. ball moving with velocity 40. If the coefficient of restitution be $\frac{1}{2}$, find the subsequent motion of each ball.

13. A fire-engine pump is provided with a nozzle the sectional area of which is 1 sq. inch and the water is projected through the

nozzle with a velocity of 130 feet per second. Find the H. P. of the engine required to drive the pump, irrespective of the loss by resistance of the working parts. A cubic foot of water weighs $62\frac{1}{2}$ lbs.

14. What is the H. P. of an engine which draws a train at a uniform rate of 45 miles per hour, against a resistance equal to 900 lbs. weight?

15. A shaft 560 feet deep and 5 feet in diameter is full of water; how many foot-pounds of work are done in emptying it, and how long would it take an engine of $3\frac{1}{2}$ H. P. to do the work? ($\pi = 3\frac{1}{7}$).

16. A mass of M pounds is drawn from rest up a smooth inclined plane of height h and length l by means of a string passing over the top of the plane and supporting a mass of m pounds hanging freely. Prove that M will just reach the top of the plane if m is detached after it has descended a distance $\frac{M+m}{m} \cdot \frac{hl}{h+l}$.

17. A particle is suspended by a string, the upper end of which is attached to a smooth vertical wall: the string is pulled aside through an angle 60° in a plane perpendicular to the wall; it is then let go, the particle impinges on the wall, and rebounds through 45°. Find the coefficient of restitution.

CHAPTER IV.

Motion of an Extended Body.

§ 1. Centre of Mass of a System of Particles.

Let A, B, C, D be four points moving with any velocities v_1, v_2, v_3, v_4.

Then if a divides AB so that

$$P \,.\, Aa = Q \,.\, aB,$$

we have shown (Chap. II, § 10) that the velocity of a is compounded of

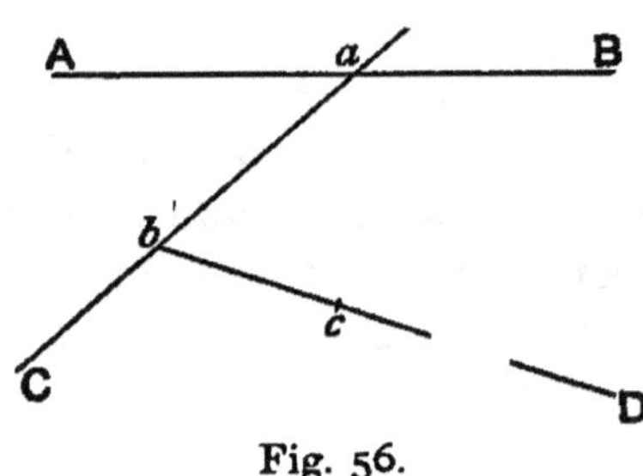

Fig. 56.

$$\frac{Pv_1}{P+Q} \text{ and } \frac{Qv_2}{P+Q}.$$

Denote this velocity by T and divide aC in b so that

$$(P+Q)\, ab = R \,.\, bC.$$

The velocity of b is then compounded of

$$\frac{P+Q}{P+Q+R} T \text{ and } \frac{Rv_3}{P+Q+R},$$

or (resolving T into its original components) of

$$\frac{Pv_1}{P+Q+R}, \quad \frac{Qv_2}{P+Q+R}, \quad \frac{Rv_3}{P+Q+R}.$$

Let U denote the velocity of b and take c in bD such that

$$(P+Q+R)\, bc = S \,.\, cD.$$

Then the velocity of c is compounded of

$$\frac{P+Q+R}{P+Q+R+S}U \text{ and } \frac{Sv_4}{P+Q+R+S},$$

or (resolving U into its components) of

$$\frac{Pv_1}{P+Q+R+S}, \quad \frac{Qv_2}{P+Q+R+S},$$

$$\frac{Rv_3}{P+Q+R+S}, \quad \frac{Sv_4}{P+Q+R+S}.$$

Now let there be particles of mass P, Q, R at A, B, C respectively. The velocity of the point b is obtained by compounding the momenta of these particles, and dividing the resultant momentum by the moving mass. b is thus related to the particles P, Q, R, as a is to the particles P, Q, and it is therefore called their centre of mass.

Similarly, if P, Q, R, S be the masses of particles at A, B, C, D, the velocity of c is obtained by compounding the momenta of the moving particles and dividing by their mass. Therefore c is called the centre of mass of this system.

§ 2. Properties of the Centre of Mass.

By the process illustrated above for two, three or four particles, we can find the centre of mass of any number of particles, defined by the following properties:—

(1) Its position can be determined when the masses and positions of the particles are known.

(2) Its velocity is represented in magnitude and direction by $\frac{\mu}{M}$, where μ is the resultant of the momenta of the particles, and M is their total mass.

In the most general case, the momentum of each particle is due partly to external forces, partly to the action of other particles; but since the momenta communicated to two particles by the stress between them are equal in magnitude and opposite in direction, the resultant momentum due to the internal forces of the system is zero. Hence the resultant momentum is compounded of the momenta due to external forces, and does not depend on the mutual actions of the particles.

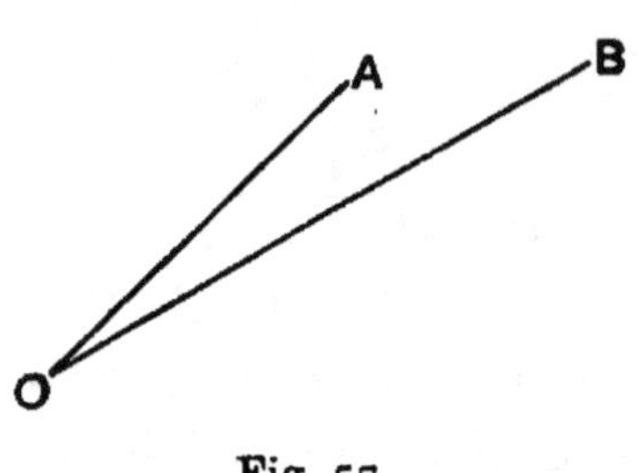

Fig. 57.

If OA, OB represent the resultant momentum of the system at the beginning and end of a very short time t, $\frac{AB}{t}$ is the rate of change of momentum, and also (Law II) represents the resultant which the external forces would have if they acted at a point.

And since $\frac{OA}{M}$, $\frac{OB}{M}$ are the velocities of the centre of mass at the beginning and end of the time t, $\frac{AB}{M.t}$ is the acceleration of the centre of mass.

Since the external forces acting at a point would communicate this acceleration to a particle of mass M, the second property of the centre of mass may be stated thus:—

The acceleration of the centre of mass of the system is that which the external forces acting at a point would impress on a particle of mass M.

Hence we may state the First and second Laws of Motion as follows:—

(1) The centre of mass of a system of particles remains at rest or in uniform motion in a straight line, except in so far as it is made to change that state by force acting on the system from without.

(2) The change of momentum of the system during any time is measured by the resultant of the impulses of the external forces during that time.

A system of particles can only have one centre of mass, for there cannot be two points (determined by the method used in § 1), whose motion is the same in all displacements from rest.

Hence in a symmetrical body, the centre of symmetry is the centre of mass: for otherwise there would be at least two points, one on each side of the centre, which have equally good claims to be regarded as the centre of mass.

Therefore the middle point of a thin straight rod, and the centre of a circle or sphere, are the centres of mass of the corresponding figures.

The centre of mass of a parallelogram or cube is the intersection of the diagonals which join opposite angles.

Let A, B be the centres of mass of two parts of a body, of masses P and Q.

If M and N be the resultant momenta of the parts, the momentum of the whole body is compounded of M and N.

A E B

Fig. 58.

Divide AB in E so that $P \,.\, AE = Q \,.\, EB$.

The velocities of A and B being $\frac{M}{P}$, $\frac{N}{Q}$, the velocity of E is compounded of $\frac{M}{P} \cdot \frac{P}{P+Q}$ and $\frac{N}{Q} \cdot \frac{Q}{P+Q}$; or of $\frac{M}{P+Q}$ and $\frac{N}{P+Q}$.

Therefore E is the centre of mass of the whole body.

Conversely, if the centres of mass of the whole body and a given part be known, the centre of mass of the remainder can be determined.

For let E, A be the centres of mass of the whole and the given part.

Then if P is the mass of the latter, R the mass of the whole body, $R-P$ is the mass of the remainder, and the required centre of mass is situated at B on AE produced, where $BE = \frac{P}{R-P} \cdot AE$.

To find the Centre of Mass of a System of Particles.

Let Oy be a given straight line, and let particles of mass m_1, m_2, m_3 start from O simultaneously with velocities $y_1, y_2, y_3, \ldots$ along Oy, not necessarily all positive.

The velocity of their centre of mass is

$$\frac{m_1y_1 + m_2y_2 + m_3y_3 + \ldots}{m_1 + m_2 + m_3 + \ldots}.$$

Hence by considering the positions of the particles after unit time, we find that the centre of mass of particles m_1, m_2, $m_3, \ldots$ at distances $y_1, y_2, y_3, \ldots$ from O along Oy, is at a distance $\bar{y}$ from O, where $\bar{y} = \frac{m_1y_1 + m_2y_2 + m_3y_3 + \ldots}{m_1 + m_2 + m_3 + \ldots}$.

Draw Ox perpendicular to Oy, and let the particles start from the positions $y_1, y_2, \ldots$ simultaneously with velocities x_1, x_2, $x_3, \ldots$ parallel to Ox; then, if the velocity of the centre of mass is $\bar{x}$,

$$\bar{x} = \frac{m_1x_1 + m_2x_2 + m_3x_3 + \ldots}{m_1 + m_2 + m_3 + \ldots}.$$

Hence when the particles are at distances x_1, x_2, $x_3 \ldots$ from Oy and $y_1, y_2, y_3 \ldots$ from Ox, their centre of mass is distant $\bar{x}$ from Oy, and $\bar{y}$ from Ox.

The particles having attained this position, let them move perpendicularly to the plane xOy with velocities $z_1, z_2, z_3, \ldots$, velocities being considered positive or negative according as they are towards or away from the reader's eye. Then, if the centre of mass then moves with velocity $\bar{z}$,

$$\bar{z} = \frac{m_1 z_1 + m_2 z_2 + m_3 z_3 + \ldots}{m_1 + m_2 + m_3 + \ldots}.$$

Hence when the particles are distant $z_1, z_2, z_3, \ldots$ respectively from the plane xOy, the distance of their centre of mass from the plane is $\bar{z}$.

Centre of Mass of a Circular arc.

If a closed figure, formed of straight uniform rods, is moving so that each rod has the same velocity perpendicular to its length, the resultant momentum is zero, for the polygon of momenta is similar to the given figure. By increasing the number and diminishing the length of the sides we can extend the result to a figure bounded by any closed curve.

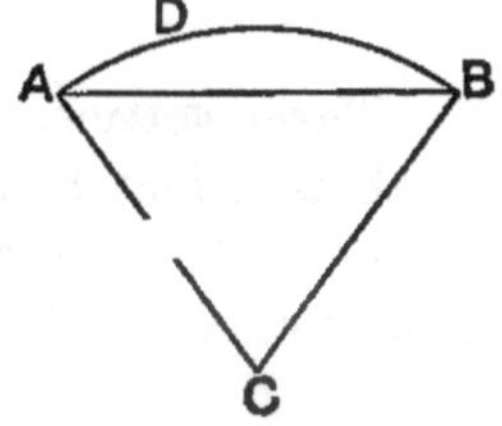

Fig. 59.

Hence we can determine the centre of mass of a circular arc ADB of radius r, subtending an angle θ at the centre C.

Let particles leave the centre with velocity r, so that after unit time they are uniformly distributed over the circle, a mass m lying on unit length of the arc, and let $AB = l = 2r \sin \frac{\theta}{2}$.

The considerations just given show that the resultant momentum is that of a mass lm moving with velocity r perpendicular to AB, i.e. it is $2mr^2 \sin \frac{\theta}{2}$.

Since the moving mass is $mr\theta$, the velocity of the centre of mass is

$$\frac{2r \sin \frac{\theta}{2}}{\theta}.$$

Therefore when the particles are at distance r from the

centre, their centre of mass is on the perpendicular from C to AB at a distance $\frac{2r}{\theta}\sin\frac{\theta}{2}$ from C.

For a semi-circle this distance is $\frac{2r}{\pi}$.

Examples.

1. Masses of 2, 4, 6, 7 lbs. lie on a square table; their distances from one edge are 2, 3, 4, 6 feet, and from an adjacent edge 3, 5, 1, 2 feet respectively. Find the positions of their centre of mass.

Here $m_1 = 2,\ m_2 = 4,\ m_3 = 6,\ m_4 = 7;$
$x_1 = 2,\ x_2 = 3,\ x_3 = 4,\ x_4 = 6;$
$y_1 = 3,\ y_2 = 5,\ y_3 = 1,\ y_4 = 2.$

Therefore $\bar{x} = \frac{4+12+24+42}{19} = \frac{82}{19};$

$$\bar{y} = \frac{6+20+6+14}{19} = \frac{46}{19}.$$

2. Three vertices of a tetrahedron are 7, 8, 9 feet above, and the fourth is 5 feet below a fixed horizontal plane. Masses of 7, 5, 12 and 15 lbs. are placed one at each vertex. Find the distance of the centre of mass from the fixed plane.

$$m_1 = 7,\ m_2 = 5,\ m_3 = 12,\ m_4 = 15;$$
$$z_1 = 7,\ z_2 = 8,\ z_3 = 9,\ z_4 = -5.$$

Therefore $(7+5+12+15)\,\bar{z} = 49+40+108-75 = 122.$

And $\bar{z} = \frac{122}{39}.$

The centre of mass is $3\frac{5}{39}$ feet above the plane.

§ 3. Moments of Inertia.

If a particle of mass m moves with angular velocity ω in a circle of radius r its velocity is $r\omega$ and its kinetic energy is $\frac{1}{2}mr^2\omega^2$.

If there are several particles $m_1, m_2, \ldots m_n$, not necessarily in the same plane, revolving with angular velocity ω about a fixed axis at distances $r_1, r_2, \ldots r_n$ from it, the kinetic energy of the system of particles is

$$\tfrac{1}{2}(m_1r_1^2 + m_2r_2^2 + \ldots + m_nr_n^2)\,\omega^2.$$

The term in brackets is called the Moment of Inertia of the system about the axis.

If a particle of mass $m_1 + m_2 + \ldots + m_n$ revolve round the same axis at a distance k from it, its kinetic energy is equal to that of the given system, if

$$(m_1 + m_2 + \ldots + m_n)\, k^2 = m_1 r_1^{\,2} + m_2 r_2^{\,2} + m_3 r_3^{\,2} + \ldots + m_n r_n^{\,2}.$$

The quantity k determined from this relation is called the Radius of Gyration of the system of particles.

Since a rigid body revolving round an axis may be considered as a system of particles similar to that which we have considered, its moment of inertia about an axis can always be determined by analysis, and in a few cases it can be determined very simply.

Moment of inertia of a thin ring about an axis through its centre perpendicular to its plane.

Let n particles each of mass m be distributed uniformly round the circumference of a ring of radius a.

The moment of inertia required is nma^2 or Ma^2 if M is the total mass of the particles.

Let the number of particles be indefinitely increased, the total mass remaining unchanged; then ultimately the mass is uniformly distributed round the ring, and the moment of inertia is still Ma^2.

Moment of inertia of a uniform thin rod AB *about an axis perpendicular to its length through its extremity* A.

Let l be the length of the rod, and divide it into n equal parts of length h, so that $nh = l$.

Let equal particles, each of mass m, be placed on the rod at distances h, $2h$, ... nh from A. nm is their total mass, which we shall denote by M.

Their moment of inertia is $m\{h^2 + (2h)^2 + \ldots + (nh)^2\}$

$$\text{or} \quad \frac{mh^2}{6} n(n+1)(2n+1),$$

$$\text{or} \quad \frac{Ml^2}{3}\left(1+\frac{1}{n}\right)\left(1+\frac{1}{2n}\right).$$

Now increasing n indefinitely, we approach the case of a uniform thin rod of mass M, and the moment of inertia approaches indefinitely to the value $\frac{Ml^2}{3}$.

Moment of inertia of a circular disc of radius a *about an axis through its centre perpendicular to its plane.*

Divide the disc into concentric rings by circles of radii $h, 2h, 3h, \ldots (n-1)h$ where $nh = a$.

The surface of the ring contained between the circles whose radii are $(r-1)h$ and rh is $\pi h^2\{r^2 - (r-1)^2\}$ which lies between $2\pi h^2(r-1)$ and $2\pi h^2 r$.

The moment of inertia of this ring lies between

$$2\pi h^2(r-1).(r-1)^2 h^2\sigma \quad \text{and} \quad 2\pi h^2 . r . r^2 h^2 \sigma,$$

σ being the quantity of matter on unit area of the disc.

The moment of inertia of the disc lies between

$$2\pi h^4\sigma(0^3 + 1^3 + 2^3 + \ldots + (n-1)^3),$$

$$\text{and} \quad 2\pi h^4\sigma(1^3 + 2^3 + \ldots + n^3)$$

$$\text{or} \quad 2\pi\sigma h^4 \frac{n^2(n-1)^2}{4} \quad \text{and} \quad 2\pi\sigma h^4 \frac{n^2(n+1)^2}{4}.$$

Now when n is indefinitely increased, these quantities approach indefinitely to $\frac{1}{2}\pi\sigma a^4$ since $nh = a$.

And $\pi\sigma a^2 = M$ the mass of the disc.

Therefore the moment of inertia of the disc about an axis through its centre perpendicular to its plane is $\frac{1}{2}Ma^2$.

The kinetic energy of a material system of mass m *is equal to the kinetic energy of a mass* m *moving with the velocity of the centre of mass of the system, together with the kinetic energy due to the motion of the parts of the system relatively to their centre of mass.*

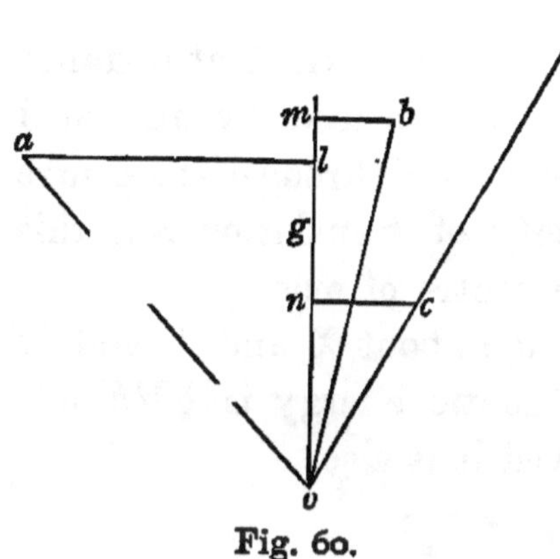

Fig. 60.

Let A, B, C,... be the masses of the particles which form the system, oa, ob, oc, ... their velocities, not necessarily all in the same plane, og the velocity of the centre of mass of the system.

Draw al, bm, cn ... perpendicular to og.

Distances on og are to be considered positive or negative according as they are measured along og or go.

The velocity of G is og; by § 1 it is also

$$\frac{A \,.\, ol + B \,.\, om + C \,.\, on + \ldots}{A + B + C + \ldots}$$

$$= \frac{A(og + gl) + B(og + gm) + C(og + gn) + \ldots}{A + B + C + \ldots}$$

$$= og + \frac{A \,.\, gl + B \,.\, gm + C \,.\, gn + \ldots}{A + B + C + \ldots}.$$

Therefore $A \,.\, gl + B \,.\, gm + C \,.\, gn + \ldots = 0.$

Now
$$oa^2 = og^2 + ga^2 + 2\, og \,.\, gl,$$
$$ob^2 = og^2 + gb^2 + 2\, og \,.\, gm,$$
$$oc^2 = og^2 + gc^2 + 2\, og \,.\, gn.$$
$$\cdot \quad \cdot \quad \cdot \quad \cdot \quad \cdot \quad \cdot \quad \cdot \quad \cdot \quad \cdot \quad \cdot \quad \cdot$$

Therefore

$$\tfrac{1}{2}A \,.\, oa^2 + \tfrac{1}{2}B \,.\, ob^2 + \tfrac{1}{2}C \,.\, oc^2 + \ldots$$
$$= \tfrac{1}{2}(A + B + C + \ldots)\, og^2 + \tfrac{1}{2}A \,.\, ga^2 + \tfrac{1}{2}B \,.\, gb^2 + \tfrac{1}{2}C \,.\, gc^2 + \ldots.$$

But ga, gb, gc, ... are the velocities of the particles relatively to the centre of mass. Therefore the proposition is proved.

In Chap I, § 5, it has been shown that an angular displacement about a given axis can be replaced by an equal angular displacement about a parallel axis together with a displacement of translation.

Hence an angular velocity ω about an axis X at distance h from the centre of mass can be replaced by an equal angular velocity about a parallel axis Y through the centre of mass, together with a velocity of translation $h\omega$, this being the actual velocity of the centre of mass.

If K, k are the radii of gyration about X and Y, and M is the mass of the body, the kinetic energy is $\frac{1}{2}MK^2\omega^2$; but by the proposition just proved it is also

$$\tfrac{1}{2} Mh^2\omega^2 + \tfrac{1}{2} Mk^2\omega^2.$$

Therefore $\frac{1}{2} MK^2\omega^2 = \frac{1}{2} M(h^2 + k^2)\,\omega^2$,

or $$K^2 = h^2 + k^2.$$

Hence the radii of gyration about all parallel axes equidistant from the centre of mass are the same, and the radius of gyration about an axis through the centre of mass is less than that about any parallel axis.

§ 4. Motion of an extended rigid body.

Each force acting on the body is applied to a particular particle and can be resolved into components acting on the particle.

Let the lines of action of the forces be parallel to a fixed plane X, the motion also being parallel to this plane.

Then any displacement is a motion of translation, or a motion of rotation about an axis perpendicular to the plane X.

Let the body have a small displacement of rotation through an angle θ, round an axis through O perpendicular to X, the plane of the paper; and let A, at which a force P is applied, be displaced to B.

Then if $OA = r$, arc $AB = r\theta$, and the arc can be regarded as a straight line since θ is very small.

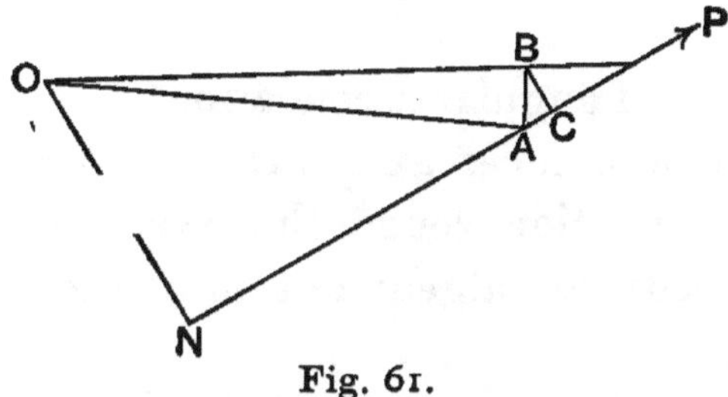

Fig. 61.

Draw ON, BC perpendicular to P.

Then $P \,.\, AC$ is the work done by P in the displacement.

But since OAN, ABC are similar triangles,

$$\frac{AC}{ON} = \frac{AB}{OA} = \theta.$$

Therefore $P.AC = P.ON.\theta$.

$P.ON$ is called the moment of P round O, or, more strictly, round the axis through O perpendicular to the plane X.

Thus the work done by P during the rotation is the moment of P multiplied by the angle of rotation.

Since work would be done against P if the direction of P were reversed, we regard $P.ON$ as positive or negative, according as the force P tends to help or hinder the rotation.

Let P be replaced by components Q and R acting at A in the plane X.

The work done by Q and R in the rotation θ is equal to the work done by P (Chap. III, § 2).

Hence the sum of the moments of two or more forces about a point in this plane is equal to the moment of their resultant about the same point.

Let ω_0 and ω be the initial and final angular velocities of the body, Mk^2 its moment of inertia about the axis through O. Then $\frac{1}{2}Mk^2(\omega^2 - \omega_0{}^2)$ is the kinetic energy gained in the displacement θ, and since this is equal to the work done

$$P.ON.\theta = \tfrac{1}{2} Mk^2 (\omega^2 - \omega_0^2),$$

if P is the only force acting.

But $\omega^2 - \omega_0^2 = 2a\theta$, where a is the angular acceleration. Therefore $P.ON = Mk^2a$.

Mk^2a is called the moment of angular acceleration.

Thus the moment of a force round an axis is the measure of its tendency to produce rotation round the axis, and if several forces act on the body the algebraic sum of their moments is equal to Mk^2a.

Since the conditions of equilibrium under forces in one plane are satisfied when the centre of mass is at rest, and the angular acceleration round any axis vanishes, they may be stated as follows—

(1) The external forces acting on the body are parallel and proportional to the sides of a closed polygon taken in order.

(2) The algebraic sum of the moments of all the forces round any point in their plane is zero.

§ 5. Resultant of Parallel Forces.

Let P, Q be two parallel forces applied to a body at the points A and B.

It is required to find a single force R which has the same effect on the motion of the centre of mass, and the same moment round any point in the plane as P and Q together.

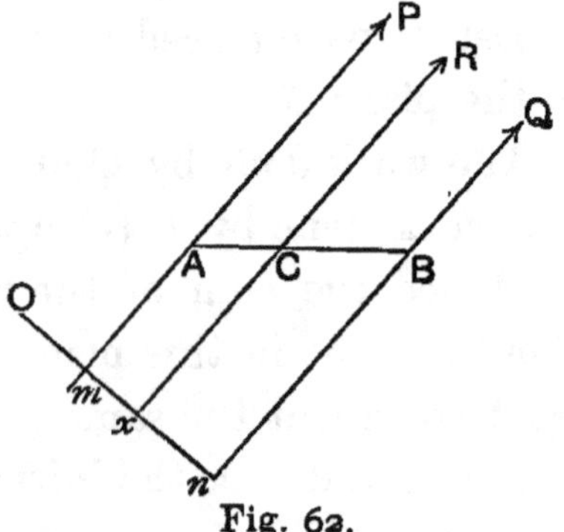

Fig. 62.

The first condition shows that R is parallel to P and Q, and equal to $P+Q$, P and Q having opposite signs if they are oppositely directed.

From any point O in the plane draw Omn perpendicular to P and Q, and let x be the point in mn through which R acts.

Then $$Om\,.\,P + On\,.\,Q = Ox\,.\,(P+Q),$$
or $$Q\,(On - Ox) = P\,(Ox - Om),$$
or $$Q\,.\,xn = P\,.\,xm.$$

Draw xC parallel to mA meeting AB in C.

Then $$\frac{P}{Q} = \frac{nx}{mx} = \frac{BC}{AC}.$$

The algebraic sum of the moments of P and Q round x is $P.\,xm - Q\,.\,xn$, which has been shown to be zero.

If the forces P and Q are oppositely directed, let P, Q denote their numerical magnitudes, P being greater than Q. Then the resultant is $P - Q$ in the same direction as P, and the point C through which it acts divides AB externally, so that

$$\frac{CA}{CB} = \frac{Q}{P}.$$

The ratio in which C divides AB does not depend on the position of O, and therefore the moment of the force R round any other point in the plane is equal to the sum of the moments of P and Q round the same point.

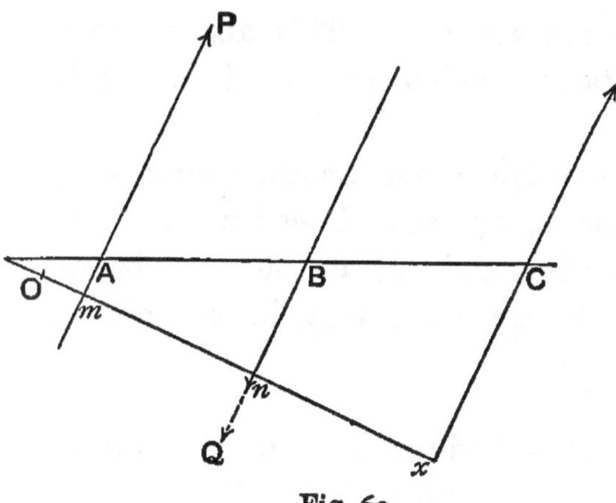

Fig. 63.

Neither does the ratio depend on the direction of P and Q. C is the same if this direction is altered.

§ 6. **Couples.**

In the case when the oppositely directed parallel forces are equal, our construction of the resultant fails. This combination of forces is called a couple ; it does not affect the motion of the centre of mass.

Let equal forces P be directed along Am and Bn.

The work done by them in a small rotation θ round O is

$$P(On - Om)\theta \quad \text{or} \quad P.mn.\theta.$$

The distance mn between the lines of action of the forces is called the arm of the couple, and $P.mn$ is called the moment of the couple.

If there are two couples whose arms are a and b and moments Pa and Qb, the condition that they should maintain equilibrium is

$$Pa + Qb = 0.$$

For when this is satisfied there is no angular acceleration, and the couples do not affect the motion of the centre of mass.

The sign of the moment is changed if the directions of the forces of the couple are reversed. This also reverses the direction of the rotation which the couple tends to produce.

Two couples therefore balance one another when they tend to produce rotations in opposite directions and have numerically equal moments; and one couple may be replaced by another with the same moment, in the same or in a parallel plane.

§ 7. Let us now consider a body which rotates round an axis under forces which are perpendicular to the axis but not all in the same plane.

Then it is still true, as may be seen from the proof in

§ 4, that the moment of the forces is proportional to the angular acceleration.

Let DE be the axis, and let two parallel forces P and Q act through A and B, perpendicular to the plane of the paper.

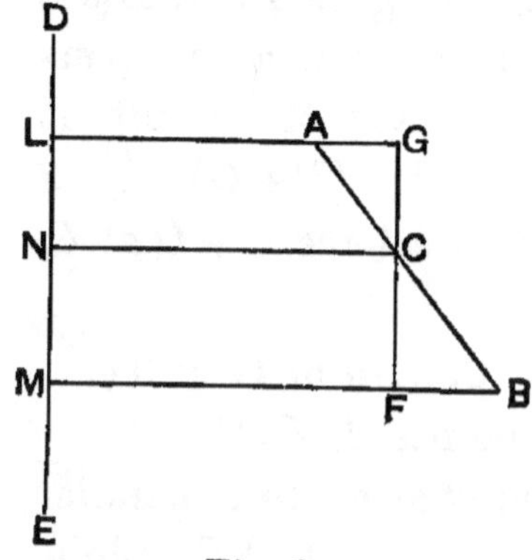

Fig. 64.

Divide AB in C so that $P.AC = Q.CB$, and draw AL, CN, BM perpendicular to DE.

Then $P.AL$ and $Q.BM$ are the moments of P and Q round DE.

Through C draw GCF parallel to DE.

Then $P.AL + Q.BM = P(GL - AG) + Q(BF + FM)$

$= (P+Q)CN - P.AG + Q.BF.$

But $$\frac{P}{Q} = \frac{CB}{AC} = \frac{BF}{AG}.$$

Therefore $$P.AG = Q.BF.$$

And $$P.AL + Q.BM = (P+Q)CN.$$

Therefore the resultant of P and Q, previously found, has the same moment round DE that P and Q together have.

Extended definition of the moment of a force.

If the force P makes an angle a with the axis, resolve it into components, $P\cos a$ parallel to the axis, and $P\sin a$ perpendicular to the axis.

The former does no work in a rotation θ round the axis, and the latter does work $Pp\sin a.\theta$ where p is the perpendicular distance of the component $P\sin a$ from the axis.

Hence generally the moment of a force round an axis is obtained by resolving the force into components at its point of application parallel and perpendicular to the axis and multiplying the latter component by its distance from the axis.

We can now show that round any axis the moment of

two parallel forces is equal to the moment of the force which we have already obtained as their resultant.

For if two parallel forces P and Q acting at A and B are inclined at an angle a to the axis, their components perpendicular to the axis are $P \sin a$, $Q \sin a$, and the resultant of these has been already found to be $(P+Q) \sin a$ acting at C; but this is the moment of a force $P+Q$ at C, acting parallel to P and Q.

§ 8. **Centre of Parallel Forces. Centre of Gravity.**

Let parallel forces P, Q, R act at points A, B, C.

By § 5 the forces P and Q are equivalent to a parallel force $P+Q$ acting at F, where $P . AF = Q . FB$. Join CF and divide it in G so that $(P+Q)\,GF = R . GC$.

Then the forces P, Q, R can be replaced by a single force $P+Q+R$ acting at G, and the position of G does not depend on the direction of the forces.

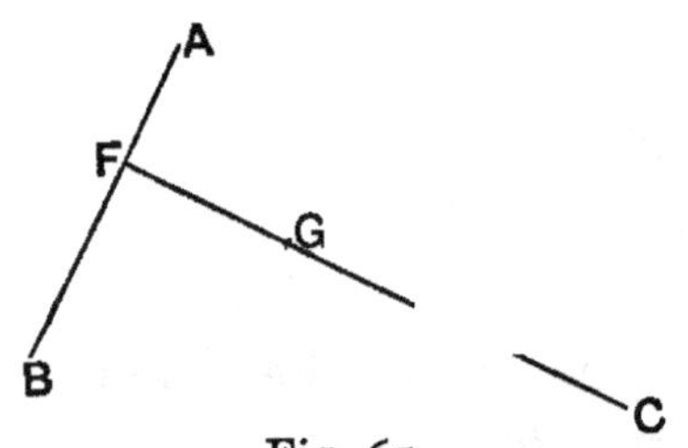

Fig. 65.

In like manner it can be shown that when several parallel forces P, Q, R... act at given points A, B, C..., their resultant passes through a certain point whose position does not depend on the direction of the forces; this point is called the Centre of Parallel Forces.

Now let $P = m_1 g$, $Q = m_2 g$, $R = m_3 g$.

Then the point G is the centre of mass of particles m_1, m_2, m_3 placed at A, B, and C respectively, and the forces are the weights of these particles.

Therefore the resultant of the weights of the particles passes through their centre of mass, and the process of finding the point through which the resultant of the weights of any number of particles acts is precisely the same as that of finding the centre of mass.

Hence the action of gravity on any body reduces to a single force, termed the weight of the body, the line of action of which always passes through the centre of mass. From this property the centre of mass is sometimes termed the centre of gravity.

If a body is supported at its centre of mass G it has no angular acceleration, for the moment of its weight round any axis through G is zero. Hence the centre of gravity of a body is sometimes defined as the point about which the body, if supported there, will balance in all positions.

The forces of gravity acting on different parts of a body cannot be considered parallel when the dimensions of the body are comparable with its distance from the Earth's centre. Here two cases arise according as the forces reduce to a single force acting through a fixed point in the body, or, not. In the former case the body is called *centrobaric*; in the latter, it cannot be said to have a centre of gravity.

§ 9. To find the centre of mass of a triangle ABC.

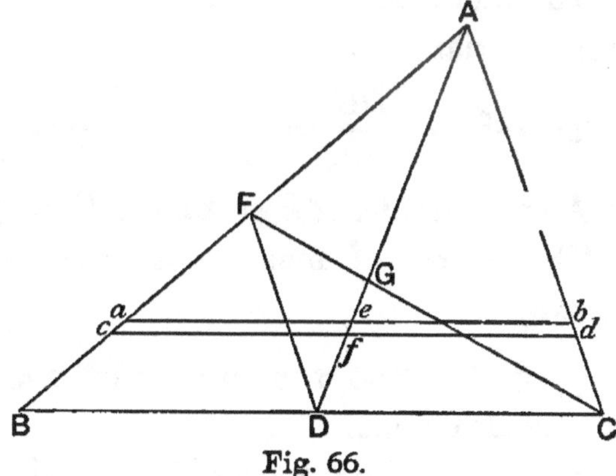

Fig. 66.

Divide the triangle into thin laminae by straight lines parallel to BC, such as ab, cd.

Bisect BC in D, and join AD, meeting ab and cd in e and f.

Then in the similar triangles aeA, BDA

$$\frac{BD}{DA} = \frac{ae}{eA},$$

similarly $\frac{CD}{DA} = \frac{be}{eA}$.

But $BD = CD$;

therefore $ae = eb$.

Similarly $cf = fd$.

Now ef being very small the lamina $abdc$ is a uniform thin rod, and its centre of mass is at the middle point of ef.

Similarly the centre of mass of every lamina formed by parallels to BC lies on AD.

Therefore the centre of mass of the triangle lies on AD.

But if BA is bisected in F, it can be similarly shewn that the centre of mass lies on CF; and therefore G, the intersection of AD and CF, is the centre of mass. Join FD; then $BF = FA$ and $BD = DC$.

Therefore FD is parallel to AC, and DFG, ACG are similar triangles.

Therefore $\frac{DG}{GA} = \frac{DF}{AC} = \frac{BF}{BA} = \frac{1}{2}$.

And G divides AB so that $DG = \frac{1}{3} AD$.

The centre of mass of a body is frequently called its centroid.

§ 10. To find the centre of mass of a pyramid on a triangular base.

Let $OABC$ be a pyramid on a triangular base ABC.

Divide the pyramid into thin triangular laminae by planes parallel to ABC, such as abc.

Bisect BC in D. Join AD and take G in AD such that $DG = \frac{1}{3} DA$.

G is the centroid of the triangle ABC.

Join OD, meeting bc in d, and OG, meeting ad in g.

Then because dga, DGA are parallel,

$$\frac{dg}{DG}=\frac{Og}{OG}=\frac{ga}{GA};$$

$$\text{but}\quad DG=\tfrac{1}{2}\,GA.$$

$$\text{Therefore}\quad dg=\tfrac{1}{2}\,ga.$$

Also because bc, BC are parallel,

$$\frac{bd}{BD}=\frac{Od}{OD}=\frac{dc}{DC},$$

$$\text{Therefore}\quad bd=dc.$$

Therefore g is the centroid of the triangle abc.

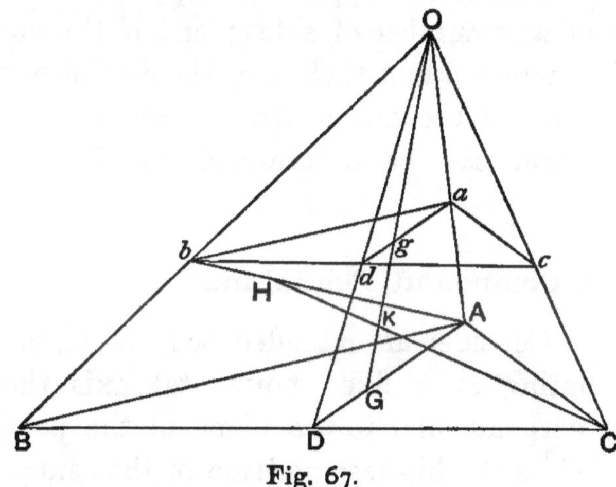

Fig. 67.

Hence when the laminae into which the pyramid is divided are very thin, their centroids all lie on OG.

Therefore the centroid of the pyramid lies on OG.

Again, regard OAB as the base of the pyramid and C as its vertex, and let b now represent the middle point of OB.

Then $bD=\frac{1}{2}\,OC$.

Take H, the centroid of the triangle OAB, and join CH.

It can be proved, as above, that the centroid of the pyramid lies on CH. And $AH=\frac{2}{3}\,Ab$, and $AG=\frac{2}{3}\,GD$.

Therefore HG, bD are parallel, and $HG = \frac{2}{3}\,bD = \frac{1}{3}\,OC$.

But since bD is parallel to OC, HG and OC are parallel, and CH and OG intersect in K, which must be the centroid of the pyramid.

And since $\dfrac{GK}{HG} = \dfrac{KO}{OC}$, $GK = \frac{1}{3}\,KO = \frac{1}{4}\,OG$.

Thus the centroid of the pyramid is found as follows :—

Find the centroid G of the base, join it to the vertex O, and in OG take a point K such that $OK = \frac{3}{4}\,OG$. K is the centroid required.

The same construction applies to a pyramid on a polygonal base of any number of sides; and if the number of the sides be increased and their lengths diminished indefinitely, the pyramid becomes a cone on any curve as base, and its centroid can be determined by the same construction.

§ 11. The Compound Pendulum.

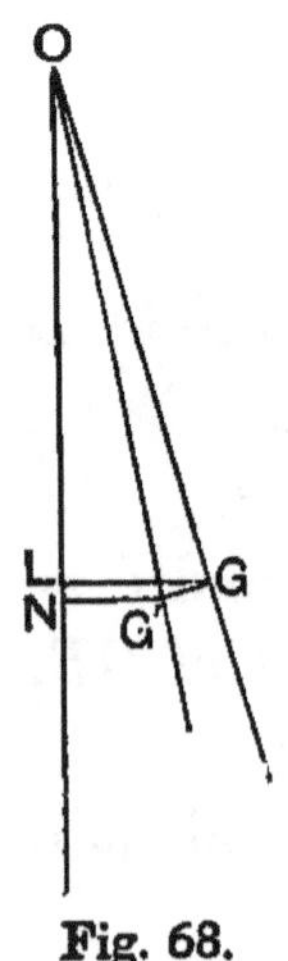

Fig. 68.

Consider an extended body of mass M oscillating on a fixed horizontal axis through O perpendicular to the plane of the paper. Let G be the highest position of the centre of mass of the body during the oscillation, G' any other position.

Draw OLN, the vertical through O, and GL, $G'N$ perpendicular to OLN. Let $GOL = \phi$, $G'OL = \theta$, $OG = h$.

Then the work done by gravity in the fall of the pendulum through the angle GOG' is

$$\begin{aligned} Mg\,.\,LN &= Mg\,(ON - OL) \\ &= Mgh\,(\cos\theta - \cos\phi). \end{aligned}$$

The resistance of the air is neglected. The only other

force acting on the pendulum is the resistance at the axis, and no work is done by this force, for it is applied at a fixed point.

The kinetic energy generated in the fall is $\frac{1}{2} MK^2 \omega^2$, where K is the radius of gyration round the axis through O, and ω is the angular velocity when the centre of mass passes through G'.

Therefore $\frac{1}{2} MK^2 \omega^2 = Mgh (\cos\theta - \cos\phi)$. (1)

Now let a simple pendulum of length l oscillate through the same angle as the compound pendulum. Its radius of gyration is l, and therefore by (1) its angular velocity ω', when it makes an angle θ with the vertical, is given by the equation

$$\frac{1}{2} l \omega'^2 = g (\cos\theta - \cos\phi).$$

Hence $\omega' = \omega$ if $\frac{K^2}{h} = l$, and the motion of the compound pendulum is precisely the same as that of a simple pendulum of length $\frac{K^2}{h}$ oscillating through the same angle, provided that the resistance of the air can be neglected.

Therefore the time of a small oscillation of the compound pendulum is $2\pi\sqrt{\frac{K^2}{hg}}$, and $\frac{K^2}{h}$ is called the length of the equivalent simple pendulum.

If k be the radius of gyration about a parallel axis through the centre of mass, $K^2 = h^2 + k^2$.

Hence the time of vibration of the compound pendulum about the axis through O is

$$2\pi\sqrt{\frac{h^2 + k^2}{hg}},$$

and it is the same about all parallel axes equidistant from the centre of mass.

Axes of Suspension and Oscillation.

We shall now show that there are two different values of h, for which the period of oscillation is the same.

If $l = \frac{h^2+k^2}{h}$, l is the length of the equivalent simple pendulum, and, when this is given, h is determined by the equation

$$h^2 - lh + k^2 = 0.$$

If h_1, h_2 be the roots of this equation,

$$h_1 + h_2 = l, \quad \text{and} \quad h_1 h_2 = k^2.$$

Therefore the times of vibration about parallel axes, at distances h_1, h_2 from the centre of mass, are equal.

The axis on which the pendulum swings is called the Axis of Suspension, and the parallel axis distant $h_1 + h_2$ from it, about which the time of vibration is the same, is called the Axis of Oscillation. The time of vibration about either of these axes is equal to that of the simple pendulum, whose length is the distance between them. This furnishes the most satisfactory method of determining g.

Kater's Pendulum. This was designed by Captain Kater in 1818.

As used in the Clarendon Laboratory at Oxford it consists of a brass rod, about 40 inches long, $\frac{5}{8}$ inch broad, and $\frac{1}{8}$ inch thick, terminating at one end in a heavy metal bob.

Each of the two axes on which the pendulum can be swung is a knife edge, formed by the intersecting faces of a triangular steel prism. One prism is fixed to the bob of the pendulum, the other is mounted on a small slider, which can be clamped anywhere on the rod.

The rod also carries another slider, which can be moved

along the rod and fixed at any desired point. By the motion of this slider the times of vibration about the knife edges can be slightly altered.

The moveable knife edge is adjusted till the times of vibration about both knife edges are nearly equal; then the slider is adjusted to make the equality still more perfect.

The time of vibration t, and the distance l between the knife edges, being accurately determined, g can be found from the formula $g = \frac{4\pi^2 l}{t^2}$.

§ 12. Atwood's Machine.

We shall now apply the principle of energy to correct the formula previously obtained for Atwood's Machine.

Let m, m' be the masses ($m > m'$), r the radius of the pulley, k its radius of gyration, and μ its mass.

Then if v is the velocity of the descending mass after a fall h, v is also the velocity of the rim of the pulley, and $\frac{v}{r}$ is the angular velocity of the pulley.

Since $\mu \frac{k^2}{2} \frac{v^2}{r^2}$ is the energy of the pulley, and $\frac{1}{2}(m + m')v^2$ is the energy of the masses m, m', the total energy is $\frac{1}{2}(m + m')v^2 + \mu \frac{k^2 v^2}{2r^2}$.

The work done in the fall h is $(m - m')gh$.

Therefore
$$v^2 = \frac{2(m - m')gh}{m + m' + \frac{\mu k^2}{r^2}}.$$

And the acceleration is
$$\frac{(m - m')g}{m + m' + \frac{\mu k^2}{r^2}}.$$

$\frac{\mu k^2}{r^2}$ can be calculated when the mass form and dimensions of the pulley are known.

The effect of the inertia of the pulley can be represented by supposing a mass $\frac{\mu k^2}{2 r^2}$ added to each pan.

If T and T' are the tensions in the strings from which m and m' are suspended, we have

$$T' = m' (g + a); \; T = m (g - a).$$

And $T' - T = (m' - m) g + a (m + m') = \frac{\mu k^2 g (m - m')}{(m + m') r^2 + \mu k^2}$.

§ 13. **Impulse.**

It has been shown that if a force F acts on a body which is free to turn round a fixed axis perpendicular to F,

$$p F = M k^2 a,$$

p being the distance of F from the axis, M, k the mass and the radius of gyration of the body, a the angular acceleration.

If the force F acts during a time t, $Ft = I$, the impulse of F; and $a t = \omega - \omega_0$, ω_0 and ω being the angular velocities at the beginning and end of the time t.

Therefore $p I = p F t = M k^2 a t = M k^2 (\omega - \omega_0)$.

When ω is the angular velocity, $M k^2 \omega$ is called the moment of angular momentum.

Example.—The ballistic pendulum.

(*a*) A shot m is fired horizontally with velocity v into a heavy pendulum of mass M whose moment of inertia round its axis of suspension is k. To find the motion, supposing that the axis is smooth.

Let I be the impulse of the blow applied at a point distant p from the axis, ω the angular velocity immediately after impact.

Then $Mk^2 \omega = pI$.

Now by the third Law of Motion, the impulse on the shot is equal and opposite to that on the pendulum, and by the second Law,

$$m(v - p\omega) = I.$$

$$\therefore \quad (Mk^2 + mp^2)\,\omega = mpv.$$

It is supposed that the shot comes to rest in the pendulum, before the pendulum has sensibly moved from its position of rest. By making M very large this can be secured, for then ω is very small.

Hence the kinetic energy of the shot and pendulum after the blow is

$$\tfrac{1}{2}(Mk^2 + mp^2)\,\omega^2.$$

Let a be the angle through which the pendulum rises, h the distance of its centre of mass below the axis.

Then $h(1 - \cos a)$ is the height through which the centre of mass of the pendulum rises, and $p(1 - \cos a)$ is the height through which the ball rises.

Therefore $(Mh + mp)\,g\,(1 - \cos a)$ is the work done against gravity.

$$\therefore \quad 4(Mh + mp)\,g \sin^2\frac{a}{2} = (Mk^2 + mp^2)\,\omega^2$$

$$= \frac{m^2p^2v^2}{Mk^2 + mp^2}.$$

Since M, m, h, p and g can all be determined, and the angle a through which the pendulum rises can also be observed, the velocity of the shot can be found.

(*b*) To determine the impulsive pressure on the axis when the pendulum is struck by a known impulse I.

Here $Mk^2\omega = pI$.

Since the initial velocity of the centre of mass is horizontal and is $h\omega$, $Mh\omega$ is the resultant impulse on the pendulum, and is horizontal; and the impulse exerted by the axis on the pendulum is $Mh\omega - I$. Hence the impulse of the pendulum on the axis is

$$I - Mh\omega \quad \text{or} \quad I\left(1 - \frac{ph}{k^2}\right).$$

There is no impulse on the axis if $p = \dfrac{k^2}{h}$.

The point whose vertical distance from the axis is $\dfrac{k^2}{h}$ is called the Centre of Percussion.

When two bodies rest in contact along a smooth circle, the action of one on the other reduces to a single force, for the pressures at the circumference of the circle are all perpendicular to it, and their lines of action pass through the centre. Hence when a body turns on a smooth hinge, the action of the hinge on the body can be represented by a single force.

Examples.

1. A horizontal rod of mass 6 lbs. hinged at A is kept in position by a string attached at B, and making an angle $30°$ with BA. Find the tension in the string.

2. Forces act along the sides of a triangle ABC in the directions AB, BC, and AC, and are proportional to the sides. Find a point in AB about which the moment of the forces vanishes.

3. Forces act on a body which are directed along, and are proportional to, the sides of a polygon. Show that they reduce to a couple, whose moment is proportional to the area of the polygon.

4. Prove that the moment of inertia of a circular disc about a diameter is half the moment of inertia about an axis perpendicular to the disc through its centre.

5. Three equal heavy particles lie at the vertices of a triangle. Show that their centroid is the centroid of the triangle.

6. Three equal triangles are cut off, one at each corner, from a given triangle. Show that the centroid of the remaining figure is the centroid of the triangle.

7. From a uniform circular disc the portion bounded by a circle described on a radius as diameter is cut away. Find the radii of gyration of the remainder about axes perpendicular to its plane, passing (1) through the original centre of the disc, (2) the centre of gravity.

8. A rod of length l and mass m hangs vertically, being free to turn about its upper extremity, which is fixed. Find the impulse which being applied at the lower end will start the rod so that it will just reach the horizontal position.

9. A uniform heavy bar AB is free to turn about a fixed hinge at B, and the end A is attached by a string to a point C vertically above B, and such that $BC = AB$, the bar being horizontal. A weight equal to that of the bar is suspended from A. Find the tension of the string, and the magnitude and direction of the reaction at the hinge.

10. A circular disc of cast-iron, 10 inches in diameter and 1 inch thick, acts as a pulley for a cord carrying 10 lbs. at one end and 5 lbs. at the other. Find the angular velocity of the pulley and the linear velocity of the weights 50 seconds after starting from rest.

11. A cylinder 5 cm. long, 5 cm. diameter, and density 8·5, is set rotating by a couple with axis parallel to the axis of the cylinder. The moment of the couple being 10000 cm. dynes, find the angular velocity after 10 seconds.

12. Find the centroid of a conical shell on a plane base bounded by right circular cones on the same axis with equal vertical angles.

Deduce the position of the centroid of the surface of a cone.

[The volume of a cone is $\frac{1}{3}$ that of a cylinder with the same altitude and base.]

13. A rectangular lamina, whose shorter edges are 4 feet long, turns round one of its longer edges 50 times a minute. It weighs 441 lbs.; find its kinetic energy (*a*) in foot-poundals, (*b*) in foot-pounds.

14. A cylinder of uniform density whose radius is 2 feet and weight a ton revolves on its axis 150 times in a minute. Find its kinetic energy. If its motion were retarded by a tangential force of 60 poundals, how many turns would it make before coming to rest?

15. A circular lamina of uniform density revolves on an axis through its centre perpendicular to its plane. If its mass is 100 lbs., its diameter $3\frac{1}{2}$ feet, and it turns round the axis 150 times a minute, find its kinetic energy in foot-pounds.

16. A mass of 10 lbs. moves round the perimeter of a regular hexagon with a constant velocity of 100 feet a second. Find the magnitude and direction of the impulse on the mass when it passes a corner of the hexagon.

CHAPTER V.

SIMPLE MACHINES. STATICS.

§ 1. It has been shown that if forces $F_1, F_2, \ldots F_n$ in the same plane are applied to points $A_1, A_2, \ldots A_n$ of a material system, the work done by the forces in any displacement is equal to the kinetic energy generated.

If the different parts of the system are in motion with uniform velocity either of translation or rotation, no kinetic energy is generated, and no work is done by the forces.

Let the point A_1 be displaced in a very short time t through a distance the component of which, in the direction of F_1, is s_1.

The work done by F_1 in the displacement is $F_1 s_1$.

Let $s_2, s_3 \ldots s_n$, be corresponding component displacements of $A_2, A_3, \ldots A_n$, in the same time.

If the motion is uniform the total work done is zero.

$$\text{Therefore} \quad F_1 s_1 + F_2 s_2 + \ldots + F_n s_n = 0.$$

Let $s_1 = v_1 t, \ s_2 = v_2 t, \ldots s_n = v_n t.$

$$\text{Then} \quad F_1 v_1 + F_2 v_2 \ldots + F_n v_n = 0.$$

Now $F_1 v_1$ is the power exerted at A, for $\dfrac{F_1 s_1}{t}$ is the work done in unit time. Hence the total power exerted is zero.

If F_1, F_2 be the only forces whose points of application are displaced, $F_1 v_1 + F_2 v_2 = 0.$ (1)

The system may then be used as a machine, i.e. as an arrangement for producing uniform motion against a given force, without necessarily exerting actively an equal and opposite force.

Let motion take place against F_2 so that $F_2 v_2$ is negative; F_2 may be called the Resistance. The only condition for uniform motion is laid down in equation (1). Thus, if $v_1 > v_2$ we may maintain uniform motion by a force F_1 less than F_2.

We shall call F_1 the Driving Force. It has generally been known as 'The Power,' but it is best to use 'Power' only to denote the rate at which work can be done.

If $F_1 > F_2$, $v_2 > v_1$. This result is sometimes applied to obtain very rapid motion against small resistance, where it is inconvenient to apply force at a rapidly moving point.

The Machines that we shall examine are the Lever, the Pulley, the Wheel and Axle, the Inclined Plane, and the Screw.

§ 2. The Lever.

The Lever is a rod, straight or curved, constrained to turn round a fixed axis called the Fulcrum. Let ACB represent the Lever, C the point where the axis meets the plane of the forces, W the Resistance, P the Driving Force. Draw Cm, Cn perpendicular to Am, Bn, the lines of action of P and W.

If the lever is pivoted on a smooth hinge at C, no work is done by the resistance at C during displacement.

Let the weight of the lever be negligible in comparison with P and W.

The condition of uniform motion, or of equilibrium, is obtained by taking moments round C, and is

$$P \cdot Cm = W \cdot Cn.$$

Thus a Driving Force P at A can balance a Resistance W at B, and by suitably choosing the magnitudes of Cm and Cn, the ratio $\frac{W}{P}$ can be made as great or as small as we please.

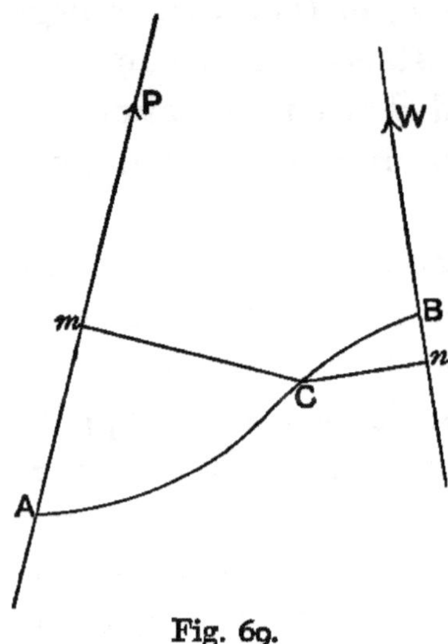

Fig. 69.

$\frac{W}{P}$ is called the Mechanical Advantage.

Levers have been divided into three classes in which the Fulcrum, the Resistance, and the Driving Force respectively occupy the middle position.

The crowbar or spade is a lever of the first kind; a pair of nutcrackers, a wheel-barrow, and an oar are levers of the second kind; a pair of tongs, used for lifting coal, is a lever of the third kind.

The relation $Pv_1 + Wv_2 = 0$, shows that, the greater the efficiency, the smaller is the ratio $\frac{v_2}{v_1}$. That is, the greater the efficiency, the slower is the motion of B when the speed of A is constant. This may be expressed by the statement—'what is gained in efficiency is lost in time.'

In many cases, the directions of P and W are parallel, and ACB is straight. Then $\frac{Cm}{Cn} = \frac{CA}{CB}$, and $\frac{CA}{CB}$ is the Mechanical Advantage.

§ 3. **The Balance.**

The Balance is a form of lever, which is employed to determine the mass of a body, by comparison of its weight with the weight of one or more known masses. As the masses of bodies are proportional to their weights, this is equivalent to a direct comparison of masses.

The form generally chosen for an accurate physical balance is roughly shown in vertical section in the figure. The lever *AB* which can turn very freely on its support is

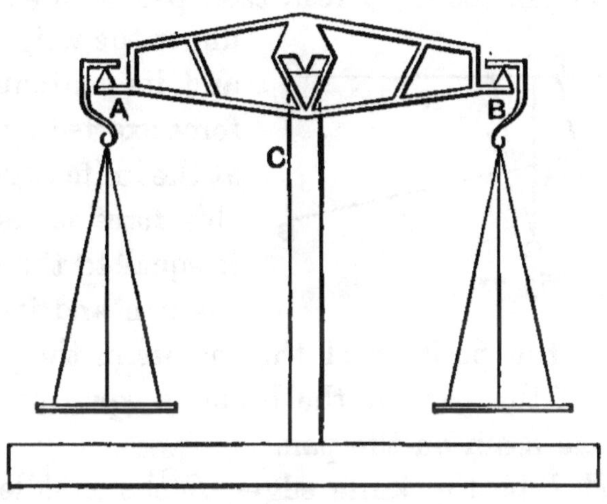

Fig. 70.

called the beam, and equal pans are hung from *A* and *B* in which the masses to be compared are placed.

In a sensitive balance the beam should be both strong and light; these advantages are obtained when it has the latticed form shown in the figure. In order that the beam may turn freely it is pivoted on the column *C* by a horizontal knife edge, which is the edge of a steel prism attached to the beam.

The pans of the balance are suspended from the beam by knife edges similar to the central knife edge.

A long pointer, which is vertical when the beam is horizontal, is attached to the beam, and its lower end moves, as the balance swings, over a graduated scale. Thus a very small displacement of the beam can be detected.

An arrangement called the Arrestment is employed for taking the weight of the beam and pans off the knife edges when the balance is not in use.

For further experimental details the reader may consult Walker's 'Theory and Use of the Physical Balance.'

When the balance is at rest, each pan is in equilibrium, under the weight of itself and its contents and the force exerted by the beam at the knife edge. Hence this force is vertical and is equal to the weight of the pan and its contents, whatever be the position of the masses in the pan. And the action of the pan on the beam is equal and opposite to that of the beam on the pan.

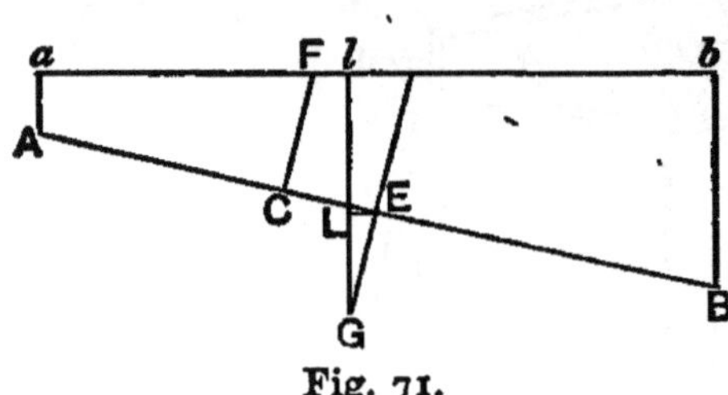

Fig. 71.

Let A, F, B be the knife edges, E the middle point of AB, G the centre of mass of the beam, and let the beam be symmetrical so that GE bisects AB at right angles.

It is required to find the position of equilibrium of the beam when masses, P and Q respectively, are placed in the pans.

Let N be the mass of the beam, M the mass of each pan. Through F draw FC perpendicular to AB, and $aFlb$ horizontal, meeting the verticals through A, G, B in a, l, b respectively.

Let $AE = a$, $CE = c$, $FC = h$, $GE = k$; $\angle EGL = \theta$, so that θ is the inclination of the beam to the horizon.

The forces which act on the beam are $(P+M)g$ at A, $(Q+M)g$ at B, Ng at G, and the resistance at the knife edge F.

Taking moments round F, and dropping the factor g

$$aF(P+M) - bF(Q+M) - Fl \, . \, N = 0.$$

Also
$$aF = (a-c)\cos\theta + h\sin\theta,$$
$$bF = (a+c)\cos\theta - h\sin\theta,$$
$$al = AE\cos\theta - EL = a\cos\theta - k\sin\theta,$$
$$Fl = al - aF = c\cos\theta - (h+k)\sin\theta.$$

Substituting the values of aF, bF, Fl,
$$(P+M)\{(a-c)\cos\theta + h\sin\theta\} =$$
$$(Q+M)\{(a+c)\cos\theta - h\sin\theta\} + N\{c\cos\theta - (h+k)\sin\theta\}.$$

Whence $\tan\theta = \dfrac{(Q+M)(a+c)-(P+M)(a-c)+Nc}{h(P+Q+2M)+N(h+k)}$.

The conditions to be satisfied by a balance are the following:—

(1) The beam should be horizontal when the pans are unloaded, i.e. $\tan\theta$ should be zero if $P = 0$ and $Q = 0$.

This requires that c should vanish; i.e. the arms of the balance should be of equal length.

If $c = 0$, we have for any load
$$\tan\theta = \frac{(Q-P)a}{h(P+Q+2M)+N(h+k)}.$$

Since $\tan\theta = 0$ when $Q = P$, the beam is also horizontal when the masses in the pans are equal.

(2) The balance should be sensitive, that is, a small difference in the masses P and Q should cause a considerable deflection of the beam.

The ratio $\dfrac{\tan\theta}{Q-P}$ is a convenient measure of the sensitiveness.

Since it is equal to $\dfrac{a}{h(P+Q+2M)+N(h+k)}$, it appears that the sensitiveness diminishes as the load increases unless $h = 0$, i.e. unless all three knife edges are in the same plane. In this case, the sensitiveness becomes $\dfrac{a}{Nk}$.

The sensitiveness is therefore increased by increasing the length of the beam, diminishing its mass, and bringing its centre of mass very near to the plane of the knife edges.

Equilibrium is altogether impossible if the centre of mass of the beam is above the plane of the knife edges.

(3) The oscillations of the beam about its position of equilibrium should be fairly rapid, for otherwise weighing is a slow process.

In determining the period of vibration of the beam, we shall suppose that $h = 0$, $c = 0$, $P = Q$.

Let α, ω be the angular acceleration and velocity of the beam; the acceleration of the extremity of the beam is then compounded of $a\alpha$ perpendicular to AB, and $a\omega^2$ along AB.

Thus if the beam only oscillates through a small angle, the acceleration of each extremity is practically $a\alpha$, upward at A and downward at B.

If the pans have no lateral swing, the forces which they exert at A and B are respectively T, T' where

$$T = (P+M)(g+a\alpha),$$
$$T' = (P+M)(g-a\alpha), \text{ since } P = Q.$$

If ϕ is the inclination of the beam to the horizontal, $\cos\phi$ is sensibly 1 since ϕ is small.

The moment of the forces which tend to increase ϕ is

$$-2(P+M)a^2\alpha - Nk\sin\phi;$$

and $Nr^2\alpha$ is the moment of the angular acceleration, where r is the radius of gyration of the beam.

Therefore $Nr^2\alpha = -2(P+M)a^2\alpha - Nk\sin\phi$,

$$\text{or}\quad \{Nr^2 + 2(P+M)a^2\}\alpha + Nk\sin\phi = 0.$$

Hence the beam oscillates like a simple pendulum of length

$$\frac{Nr^2 + 2(P+M)a^2}{Nk}.$$

In order that the oscillations may be rapid this length should be small. This condition is to some extent at variance with those for sensitiveness. The maximum sensitiveness can then only be gained at some sacrifice of rapidity in weighing.

Accurate Weighing.

When the arms of the balance are of equal length, the three knife edges in the same plane, and the beam horizontal, it may be inferred that the pans are equally loaded; but it is not easy to ensure the fulfilment of the two first conditions. There are two methods of using a balance which remove this difficulty.

I. Let a, b be the distances of the outer knife edges from the central knife edge when the pans are unloaded.

Place the mass Q which is to be determined in one pan and load the other with known masses P, till the reading of the pointer attached to the beam is the same as when the pans are unloaded.

The forces thus introduced are the weights of P and Q and an additional resistance at the central knife edge. Since the position of equilibrium is unchanged, the sum of the moments of these forces round the knife edge vanishes, and

$$Pa = Qb. \tag{1}$$

Remove the unknown mass Q from its pan and place it in the other. Balance against it known masses P' till the position of rest is the same as before.

$$\text{Then} \quad P'b = Qa. \tag{2}$$

By (1) and (2) $Q^2 = PP'$,

and since P is very nearly equal to P', we may write

$$Q = \tfrac{1}{2}(P + P').$$

II. In the left-hand pan of the balance, place a mass greater than Q. In the right-hand pan place Q together with known masses m, which bring the beam to a position of equilibrium in which the pointer falls on the scale.

Remove the contents of the right-hand pan and replace them by a mass P, which brings the beam to the same position of equilibrium. Then $Q+m$ and P have, when placed in the same pan, the same effect on the beam. Therefore they are equal, and $Q = P - m$.

The steel-yard.

Consider the case when the masses are hung directly from the beam, and one standard mass only is employed in weighing, its effect being adjusted by placing it at different points on the beam. This form of balance is called the Steel-yard.

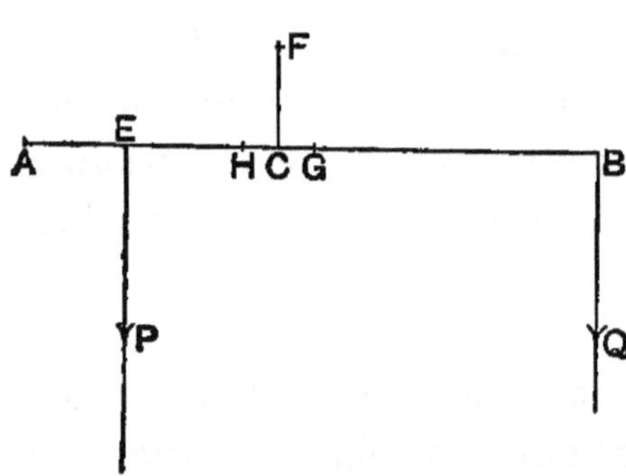

Fig. 72.

Let F be the fulcrum, G the centre of mass of the beam, P the standard mass suspended at E, Q the unknown mass suspended at B, the extremity of the beam.

Draw FC perpendicular to AB.

Let $CB = a$, $CG = c$, $EC = x$, $N =$ mass of the beam.

The beam will rest in a horizontal position if

$$Px = Nc + Qa.$$

Take a point H in CE such that $Nc = P \,.\, CH$.

Then $Qa = Px - Nc = P(CE - CH) = P \,.\, HE$.

We can therefore graduate the steel-yard by dividing

HA into parts each equal to $\frac{a}{m}$, m being any integer, and marking the scale at each division.

If the beam is horizontal when P is at the xth division from H,

$$P = \frac{Qa}{HE} = \frac{Qm}{x}, \text{ or } Q = \frac{Px}{m}.$$

Thus Q is determined at once by reading the graduations at the point where P is suspended.

§ 4. The Pulley.

The Pulley is a circular disc mounted on an axle at its centre, which revolves in a framework called the Block, and the edge of the Pulley is grooved so that a cord wound on it cannot slip off.

The Pulley is said to be Fixed or Moveable according as its Block is fixed or moveable.

In the Fixed Pulley, the Driving Force P is applied at one end of the cord and the resistance W at the other.

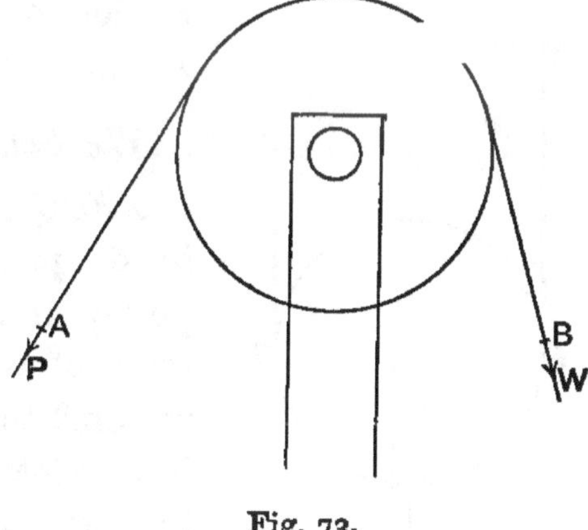

Fig. 73.

The cord between the points A and B is acted on by the forces P and W and the resistance of the pulley. The work expended by these forces is employed in bending the cord and giving it kinetic energy.

We shall suppose that the cord is very light and perfectly flexible, so that it requires no work to bend it. Under these conditions no sensible work is spent on the cord, and the moments of P and W are equal and opposite to the moment of the total pressure of the pulley on the cord. Hence the moment of the total pressure of the cord on the pulley is $a(P-W)$, where a is a radius of the pulley.

The pulley revolves under this pressure, its own weight, and the resistances at the axle. If the bearings are smooth the resistances reduce to a single force which is directed towards the axis of motion and has no moment round it. The centre of mass of the pulley is on the axis. Hence the moment of the forces which act on the pulley is $a(P-W)$. If the pulley revolves with uniform velocity no kinetic energy is generated, and $P=W$.

If the pulley revolves with angular acceleration α and has a moment of inertia K, $a(P-W)=K\alpha$.

Hence the tension of the string is not the same throughout, unless the velocity is uniform or zero.

The Fixed Pulley can only be used to change the direction of a force; its Mechanical Advantage is 1.

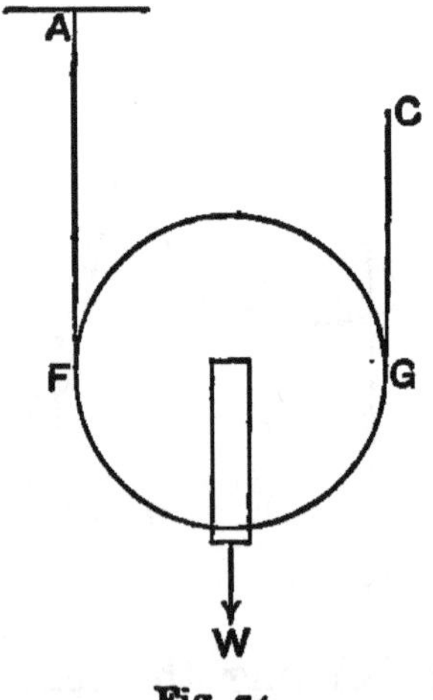

Fig. 74.

The Single Moveable Pulley.

AFGC is a cord, with the end *A* fixed, passing under a moveable pulley from the block of which a mass, of weight Q, is hung. It is required to raise the mass with uniform velocity by applying the driving force P as a tension to the string GC. Let w be the weight of the pulley. If v_1, v_2 are the vertical velocities of the block, and of a point on the string GC, $Pv_1-(Q+w)v_2=0$.

And since $$v_1 = 2v_2, \quad Q+w=2P.$$

If the weight of the pulley can be neglected, $\frac{Q}{P} = 2$; this is the Mechanical Advantage of a single moveable pulley when the strings are parallel.

The tension of the string $FA = P = \frac{1}{2}(Q + w)$.

Thus the pulley is half supported by the ceiling and half by the Driving Force.

§ 5. **First System of Pulleys.**

A greater Mechanical Advantage can be obtained by combining several pulleys. In the system shown in the figure there are two blocks, the upper fixed, the lower moveable, and several pulleys (two in the figure) may be attached to each block.

A string is attached to the upper fixed block, and passes round every pulley; the driving force P is applied to the free end of the string.

The resistance is generally the weight of a body which is raised by the lower block. Let Q be its measure in absolute units of force, and let the weight of the pulley and block be negligible. If the body is raised with uniform velocity v_2, and the point C on the string, at which P is applied, has velocity v_1,

$$P v_1 = Q v_2.$$

In the figure there are four strings at the lower block, and if the lower block rises through a distance v_2, C falls through $4\, v_2$.

Therefore $v_1 = 4\, v_2$, and $4\, P = Q$.

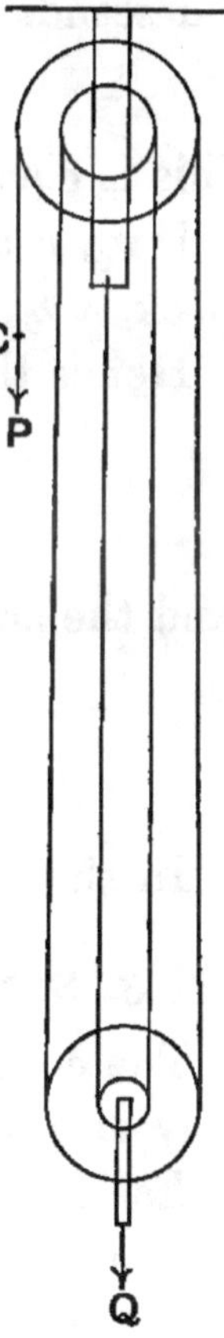

Fig. 75.

The Mechanical Advantage here is 4.

When there are n strings at the lower block the

Mechanical Advantage is n, if the weight of the lower pulley is neglected.

If the weight of the pulley is w, $Pv_1 = (Q+w)v_2$, and the Mechanical Advantage is $n - \frac{w}{P}$.

The pressure on the upper block is $P+Q+w$.

If any forces P and Q are applied to the Pulleys, it is easy to calculate the acceleration of C and of the lower block, when the masses of the pulleys can be neglected.

For when the lower block ascends through a height h, C descends through $4h$, and the work done by P and Q is

$$(4P-Q)h.$$

This is equal to the kinetic energy generated.

If v_0, v are the initial and final velocities of the lower block, $4v_0$, $4v$ will be the initial and final velocities of C.

Hence the kinetic energy generated is

$$\frac{1}{2g}(v^2-v_0^2)(Q+16P).$$

And the acceleration of the lower block is $\frac{v^2-v_0^2}{2h}$ or

$$\frac{g(4P-Q)}{Q+16P}.$$

In the general case the acceleration is $\frac{g(nP-Q)}{Q+n^2P}$.

§ 6. Second System of Pulleys.

The combination of moveable pulleys shown in the figure may also be used to obtain Mechanical Advantage.

Each pulley is supported by a string, with one end attached to the ceiling or some fixed support, and the other to the block of the neighbouring pulley above it. The Driving Force P is applied at D, and the Resistance is the weight of a body hung from the lowest block.

If D rises through a height v_1, the block K and the end of the string YC rise through $\frac{1}{2}v_1$. The end of XB rises through $\frac{1}{2}\cdot\frac{1}{2}v_1$, that of TA through $\frac{1}{2}\cdot\frac{1}{2}\cdot\frac{1}{2}v_1$; and the lowest block F rises through $\frac{1}{2}\cdot\frac{1}{2}\cdot\frac{1}{2}\cdot\frac{1}{2}v_1$ or $\frac{1}{2^4}v_1$.

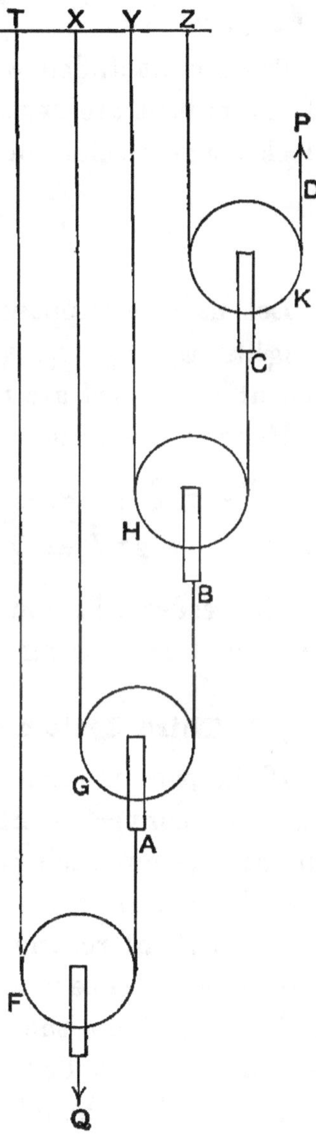

Fig. 76.

Hence if v_1 and v_2 are the velocities of the point D and the block F, the condition of uniform motion is

$$P v_1 - Q v_2 = 0,$$

assuming that the weights of the pulleys can be neglected.

And since $v_1 = 2^4 v_2$,

$$Q = 2^4 P.$$

Hence the Mechanical Advantage is 2^4 when there are four pulleys. If there are n pulleys the Mechanical Advantage is 2^n.

We may consider the block K as rising uniformly under the Force P and the tension of the cord YC. The case of the single moveable pulley shows that this tension is $2P$. In like manner the block H may be considered as rising under this tension and the tension of XB.

In this way we find that the tensions of the cords

ZD, YC, XB, TA are P, $2P$, $4P$, $8P$ respectively.

Hence the total pull on the support is $(2^4-1)P$. The reaction to this, together with P, balances Q.

When there are n pulleys the pull on the support is $(2^n-1)P$.

There is no difficulty in showing that when the weights of the pulleys are neglected the acceleration of the lowest block for any values of P and Q is

$$\frac{2^n P - Q}{2^{2n} P + Q} \cdot g.$$

Let us now suppose that the pulleys F, G, H, K have weights w_1, w_2, w_3, w_4. It has been shown that their velocities upward are v_2, $2v_2$, $4v_2$, $8v_2$.

Hence the condition for uniform motion is (§ 1),

$$Pv_1 - Qv_2 - w_1 v_2 - 2w_2 v_2 - 4w_3 v_2 - 8w_4 v_2 = 0,$$

or $$2^4 P = Q + w_1 + 2w_2 + 4w_3 + 8w_4.$$

The reader will find it a useful exercise to determine the tension of each string in this case.

§ 7. Third System of Pulleys.

If the page is turned upside down the second system of pulleys is converted into that shown in the annexed figure; in this system each pulley hangs from the block above it, and all the strings are parallel and attached to the weight.

We have interchanged the positions of the support and the resistance. Hence when there are n pulleys, $Q=(2^n-1)P$, and the pull on the support is $2^n P$, provided that the weights of the pulleys can be neglected.

The reader should have no difficulty in finding the tensions of the cords in terms of P, and thence deducing these results.

It can be easily shown that the acceleration of the

hanging body is $\frac{(2^n - 1)P - Q}{(2^n - 1)^2 P + Q} g$ for any values of P and Q.

Let w_1, w_2, w_3, w_4 be the weights of the four pulleys F, G, H, K.

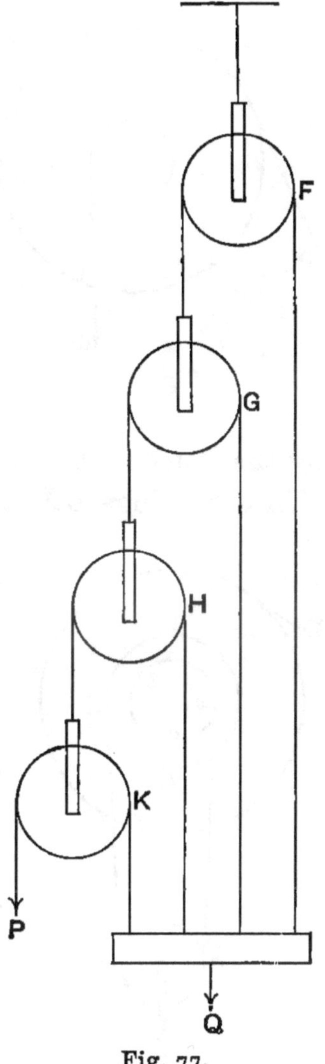

Fig. 77.

When Q rises through a height v_2, F remains at rest, G falls through v_2, H through $2v_2 + v_2$ or $3v_2$, K through $2 \times 3v_2 + v_2$ or $7v_2$, and the point of application of P through $2 \times 7v_2 + v_2$ or $15v_2$.

Hence by § 1,

$$(2^4 - 1)P + (2^3 - 1)w_4 + (2^2 - 1)w_3 + w_2 = Q.$$

§ 8. The Wheel and Axle.

This consists of two pulleys of different radii revolving in one piece on the same horizontal axis. The Resistance Q is applied to a cord coiled round the Axle, the Driving Force P to a cord wound in the opposite direction on the wheel.

If c and b are the radii of the Wheel and the Axle, the condition for motion with uniform velocity is (by taking moments)

$$Pc = Qb.$$

The Mechanical Advantage is $\frac{c}{b}$; it may be increased indefinitely by diminishing the radius of the Axle, but this tends to weaken the machine, so that it may not bear the forces applied to it.

To overcome this difficulty the Differential Axle has been used.

It consists of two co-axial circular discs ABX, abx, (Fig. 79) of radii b, b', with grooves cut in their edges, forming the same solid piece with the Wheel. One end of a cord is secured to a point on ABX; the cord quits the groove at X, passes under the moveable pulley hH, and thence passes into the groove of abx at x, the end of the string being fastened in this groove.

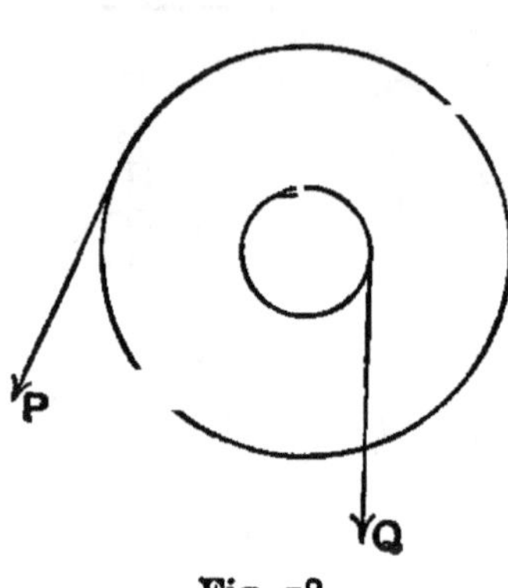

Fig. 78.

The machine is now acted on by the Force P, the resistance at the axis, and the tensions T in the strings HX, hx.

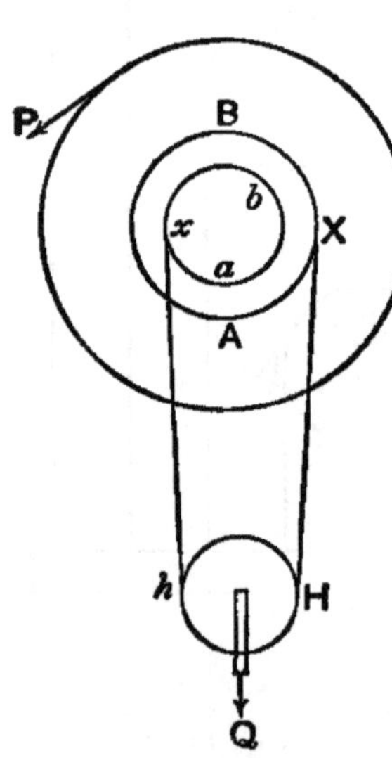

Fig. 79.

The condition for uniform motion is obtained by equating the sum of the moments round the axis to zero.

Therefore $Pc = T(b-b')$.

Also supposing that the pulley Hh has no weight,

$$Q = 2\,T\cos a\,;$$

where a is the angle which each string hx, Hx makes with the vertical.

Therefore $$P = \frac{(b-b')}{2c\cos a}\,Q.$$

§ 9. The Inclined Plane.

The Resistance is the weight of a body which is to be drawn up the plane.

Let GK be the line of greatest slope, A a particle of

weight Q resting on the plane, P the Driving Force acting up the plane.

If the plane is smooth, its action R on A is perpendicular to the plane.

Draw HL perpendicular to KG, and let GH be vertical. A is in equilibrium under forces P, Q, R, which are parallel to the sides of the triangle GHL.

Therefore $\dfrac{P}{LG}=\dfrac{Q}{GH}=\dfrac{R}{GH}$.

And $\dfrac{Q}{P}=\dfrac{GH}{LG}=\dfrac{KG}{GH}$, since GHL, GKH are similar triangles.

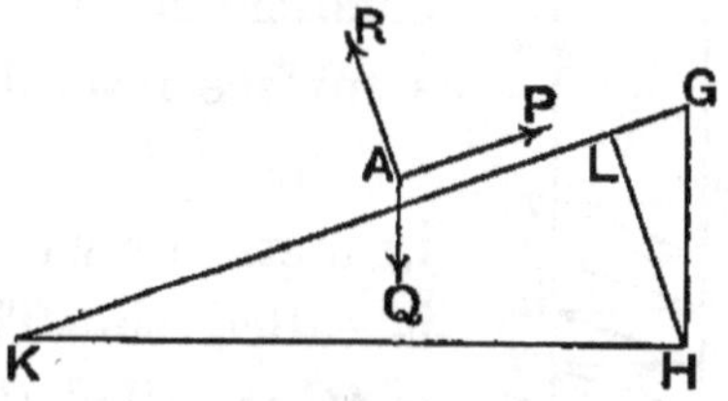

Fig. 80.

The Mechanical Advantage is therefore the ratio of the length to the height of the plane.

Next, let the Driving Force be horizontal.

In this case P is perpendicular to GH, Q to KH, and R to GK.

Therefore $$\frac{P}{GH}=\frac{Q}{HK}=\frac{R}{GK}.$$

And the Mechanical Advantage is the ratio of the base to the height of the plane.

§ 10. The Screw.

Draw two parallel lines AB, CD (Fig. 81); on AB take points $a, b, c, d, \ldots$, such that

$$Aa = ab = bc = cd = \ldots = h.$$

And draw am, bn, cp, dq, ... perpendicular to AB. Join Am, an, bp, cq, ...

If the portion of paper $ABCD$ be cut out and bent to form a cylinder with AC as the perimeter of the base, the lines AM, an, bp, cq, ... form a continuous curve on the cylinder known as the Helix.

Let a plane figure F, one side of which is the line rs ($< h$), be applied to the cylinder, so that s is at a point of the helix, rs parallel to AB, and the plane of F perpendicular to the plane which touches the cylinder along rs.

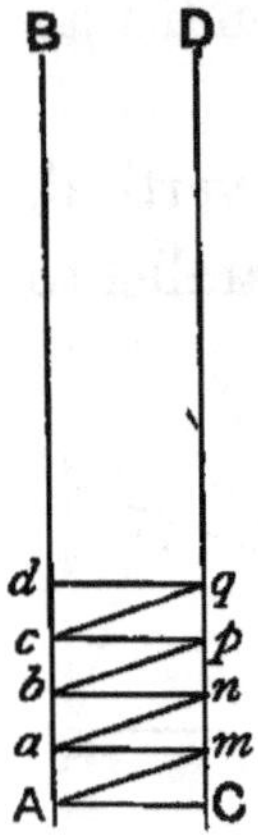

Then, if s moves along the Helix so that these conditions are satisfied, the area F traces out a surface resembling that of a corkscrew.

The cylinder and the surface when imitated in metal or wood form a screw, and the surface is called the Thread of the screw; the angle mAC is called the Pitch of the Screw.

Fig. 81.

We shall only consider the case when F is a rectangle. The screw then formed is shown in the accompanying figure. When used as a Mechanical Power the screw revolves in a fixed hollow cylinder of the same radius with its inner surface grooved to form a screw of the same pitch.

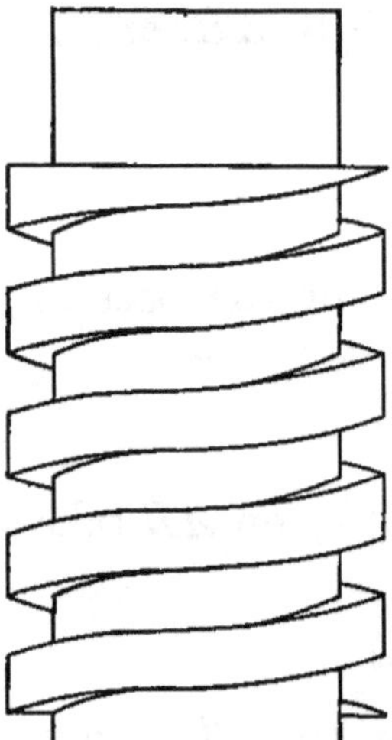

The Screw is used to obtain great pressure in the direction of its axis. The Driving Force is applied at the end of an arm perpendicular to the axis of the screw, and is itself perpendicular to the arm and to the axis.

Therefore if P is the Driving Force, and a is the length of the arm, $2\pi aP$ is the work done by P in one revolution of the arm. During this revolution the end of the screw advances by the distance h.

Fig. 82.

Therefore if W is the resistance $Wh = 2\pi aP$.

Hence the Mechanical Advantage is $\frac{2\pi a}{h}$. Since $\frac{1}{h}$ is the number of threads per unit length (inch or centimetre) of the screw, the Mechanical Advantage is obtained by multiplying the number of threads per unit length of the screw by the circumference of the arm.

§ 11. Principle of Virtual Work.

In each Machine that has been considered the conditions for uniform motion are, by the First Law of Motion, the same as the conditions for rest.

Thus in the three systems of Pulleys, a weight P applied as the Driving Force can balance weights nP, $2^n P$, or $(2^n - 1) P$, if the weights of the Pulleys are neglected.

Hence the conditions under which a material system can remain at rest in a given position may be obtained by supposing the system to be in motion through this position, and equating to zero the powers of the acting forces. Since the motion is not real but hypothetical, this method is called the Method of *Virtual* Velocities.

Example.—A smooth triangular prism ABC forms a double inclined plane, on which particles m, m′ rest, connected by a light string which passes over a pulley at A. To find the condition of equilibrium.

Let BC be horizontal, and denote the angles at B and C by B and C.

The external forces acting on the system composed of the particles and string are the resistances of the planes at P and Q, the weights mg, $m'g$, and the resistance of the pulley at A.

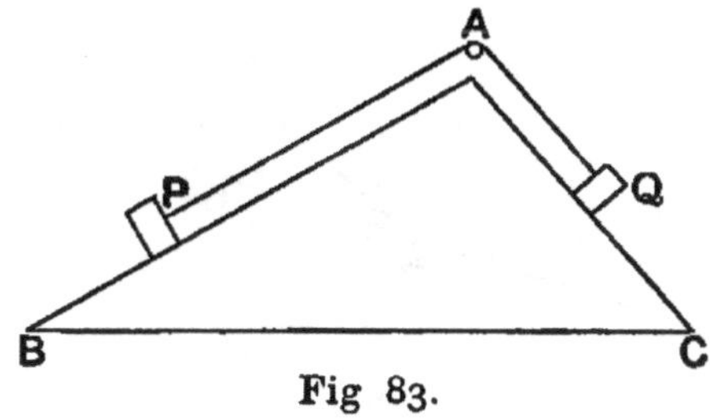

Fig 83.

The internal force is the tension T in the string.

If m moves with velocity v up AB, m' has the same velocity down AC. The tension in the string does work at the rate Tv per unit time in the displacement of m, and at the rate $-Tv$ in the displacement of m'. Therefore on the whole it does no work.

The resistances of the pulley and planes also do no work, for the motion at their points of application is perpendicular to their direction.

Therefore mg, $m'g$ are the only forces which do work, and their power is $(m'g \sin C - mg \sin B)\, v$.

Hence the condition of equilibrium is $m' \sin C = m \sin B$.

The tension in the string can be found by considering a virtual displacement of P only.

We then have $$Tv = mgv \sin B,$$
or $$T = mg \sin B.$$

In many questions it is more convenient to consider the work done in small hypothetical displacement of the points at which forces act. It has been shown in Chap. IV, § 4, that internal forces acting on a rigid body do no work in a displacement of rotation, and it also follows from Chap. III, § 2, that they do no work in a displacement of translation.

Example.—AC is the base and CB the height of a smooth inclined plane AB; D is a given point in CB produced, and two particles of mass P and R are connected by a string passing over a smooth peg at D. R is placed on AB at *Q, and P hangs freely. To find the position of equilibrium.*

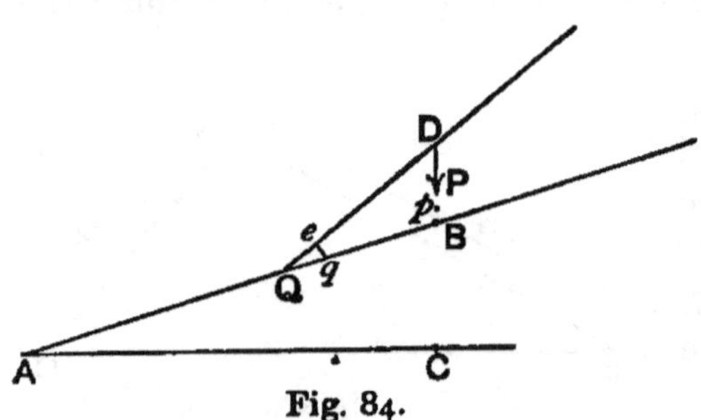

Fig. 84.

Let $DQB = \theta$, $BAC = \alpha$.

The acting forces are the weights of the particles, the resistance of the plane, and the tension of the string.

Let P be displaced downward through a distance h to p.

Then, the string being unstretched, R will be displaced up the plane to q. The tension does no work since the string remains unstretched, and the resistance of the plane does no work since it is perpendicular to the displacement at Q.

The work done by gravity on P is $Pg \, . \, h$.

Also $DQ - Dq = h$, and if qe is perpendicular to DQ, $De = Dq$ (within a negligible quantity) when h is very small.

Therefore

$$h = Qe = Qq \, . \cos \theta.$$

The vertical component of Qq is $Qq \sin a$ or $\dfrac{h \sin a}{\cos \theta}$.

Therefore the work done on R is $-Rg \, . \dfrac{h \sin a}{\cos \theta}$, and

$$Pgh - Rgh \frac{\sin a}{\cos \theta} = 0,$$

$$\text{or} \quad \cos \theta = \frac{R}{P} \sin a.$$

This method of finding the conditions of equilibrium is called the Method of Virtual Work.

§ 12. Stable and Unstable Equilibrium.

If a body is slightly displaced from a position of equilibrium X to another position Y the forces are generally slightly altered, in respect either of their magnitude or of the direction and position of their lines of action. The body therefore is generally no longer in equilibrium.

The forces which act on the body in its position Y may tend either to displace the body still further or to restore it to the position X.

In the former case the equilibrium is said to be unstable, since a slight displacement of the body from its position of rest calls into play forces which increase the displacement and prevent the body from coming to rest.

If, however, the forces tend to restore the body to the position X, they generate in the displacement from Y to X a certain kinetic energy which carries the body beyond X.

In the new displacement forces arise which either increase or check the displacement beyond X. In the former case the equilibrium is unstable, in the latter it is stable.

If in the position Y the forces which act on the body maintain it in equilibrium, the equilibrium at X is said to be neutral.

Instances of bodies in stable equilibrium are afforded by all ordinary cases of equilibrium, e.g. a stool resting on three legs, or the bob of a pendulum at rest.

Cases of unstable equilibrium are only theoretical. A cone can conceivably, but not practically, rest on its vertex, or a chair on two legs, or an egg on its end.

An instance of neutral equilibrium is the case of a sphere resting on a smooth horizontal plane.

In examining the stability of equilibrium it is useful to notice that the centre of mass of a heavy body tends to assume as low a position as it can consistently with the other constraints to which the body is subject. This position of equilibrium is stable.

Examples.

1. A cylinder with elliptic cross section rests on a smooth horizontal plane, to find the positions of equilibrium.

These are two in number, the major axis being vertical in the one, the minor axis in the other.

The first position is unstable, for gravity can do work in a displacement of the centre of mass, and the moment of the forces when the body is displaced tends to displace the body still further.

The second position is stable, since the centre of mass is in its lowest possible position, and the moment of the forces called into play by a displacement tends to restore equilibrium.

2. A prism whose section is a triangle ABC with each angle at the base AB equal to 30° rests on a rectangular face with

BC horizontal; to find the mass that can be added at A without upsetting the prism.

If the prism is slightly displaced round C the acting forces are the weight W of the prism, the weight w of the added mass at A, and the resistance at the point C.

Let G be the centroid of ABC; produce AG to D, and draw GL, AM perpendicular to BC.

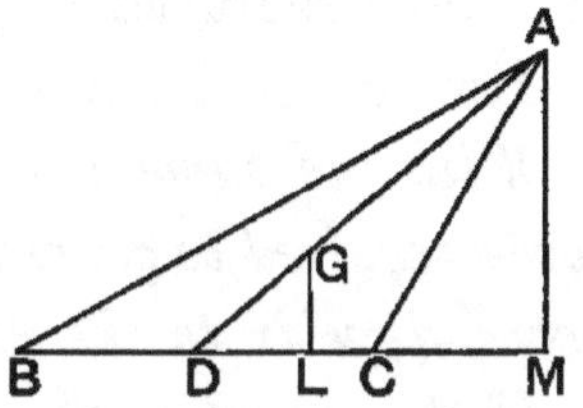

Take moments round C.

The equilibrium becomes unstable when
$$w \, . \, CM > W \, . \, CL.$$

Fig. 85.

Now $DL = \frac{1}{3} DM$, and $DC = \frac{1}{2} DM$.

Therefore
$$CL = \tfrac{1}{6} DM = \tfrac{1}{3} CM.$$

Therefore the prism will be upset when $w > \dfrac{W}{3}$.

§ 13. Friction.

We have till now assumed that when two bodies rest in contact along a plane surface, the stress between the bodies is perpendicular to this plane. Thus a body A resting on a horizontal plane could be set in motion with constantly increasing velocity by any horizontal force, however small.

Experience shows that this is not the case; small horizontal forces can be applied to A without causing it to move; equal horizontal forces opposing these must therefore exist, and they are attributed to friction between A and the surface on which it rests.

Laws of Friction.

The body A exerts a vertical force R, equal to its weight, on the plane S. If a gradually increasing horizontal force be applied to A, A will begin to move when this force attains a certain magnitude x.

Now let a body B of equal mass be placed on the top of A; the pressure on the plane is now $2R$, and experiment shows that the force required to just set A in motion is $2x$. Similarly, if A exerts a vertical pressure nR on the plane, the force required to just move A is nx.

Hence the First Law of Friction follows.

With the same surfaces in contact, the minimum tangential force required to generate motion is proportional to the pressure perpendicular to the surfaces.

If the position of A is changed so that any other part of its surface is in contact with the plane S, A exerts the same force on the plane as before, and experiment shows that the same force is required to produce motion.

Therefore the horizontal force required to generate motion is independent of the area of the surfaces in contact, the total pressure remaining the same.

This is the Second Law of Friction.

Thus if R be the force exerted by A perpendicular to S, there is generally a tangential force between A and S, which has a magnitude μR when A is on the point of motion. The Laws just enunciated imply that, when the substances of the bodies which are in contact are given, μ does not depend on their weights or dimensions. μ is a constant, and is called the coefficient of friction between these substances.

If a body slides on a rough surface under known constant forces its acceleration when small is uniform. Therefore the force of friction is constant, and the Third Law of Friction follows, viz. :—

The force of friction is independent of the velocity, when the velocity is small, and its direction is opposite to that of motion.

Friction during motion is generally somewhat less than just before motion.

Rankine supposed friction to be due to the fact that even apparently plane surfaces consist of small depressions and elevations, and that when two such surfaces are in contact their inequalities lock into one another, thus requiring a certain exertion of force to move one body over the other; and, the greater the pressure between the two surfaces, the more deeply do they interlock.

When the force of friction has attained its greatest possible magnitude, the body A is said to be in limiting equilibrium.

Angle of Friction. Cone of Friction.

If a body rests on a rough plane surface, the action of the surface on the body at any point may have any direction lying within a certain cone.

For if ON be the normal to the surface at O, make $ON = R$, the normal component of the action at O, and draw NE perpendicular to ON and equal to μR.

Fig. 86.

Then OE represents in magnitude and direction a possible limiting value of the action of the surface on the body.

The angle NOE is called the angle of friction, and the cone formed by the revolution of the triangle NOE round ON is called the cone of friction.

All straight lines drawn from O within this cone represent possible values of the action of the surface on the body, and there are no other possible values.

When a body H rests in contact with two rough surfaces P and Q, the resistance at each surface is generally indeterminate.

For let the cones of friction at P and Q be represented

in section by the angles APB, CQD. Then if any point K of the vertical KG, through the centre of mass of H, lies within both cones, forces along PK, QK, and KG can be found which maintain H in equilibrium, and the values of these forces depend on the position of K.

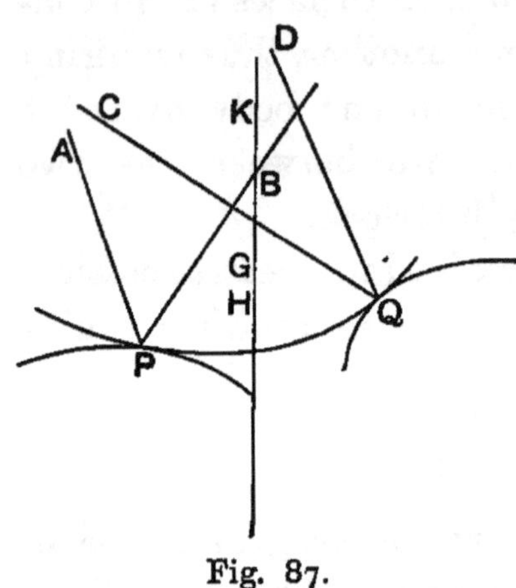

Fig. 87.

If there is limiting equilibrium at either P or Q, the problem of finding the resistances at P and Q is determinate, since then the direction of either PK or QK is known.

If no point on KG lies within both cones, the position of H is not a possible position of equilibrium.

A heavy rough particle P rests in limiting equilibrium on an inclined plane. To find the inclination of the plane.

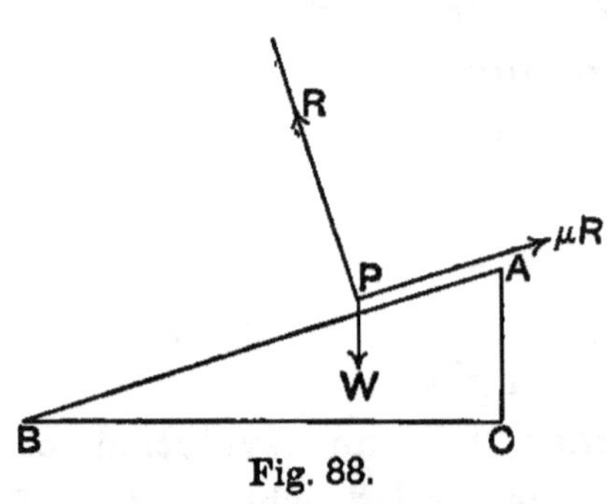

Fig. 88.

Let W be the weight of the particle P, R the action on P perpendicular to the plane; then μR is the action parallel to the plane, and is directed up the plane along the line of greatest slope.

Since the particle is in equilibrium,

$$\mu R : AC : R : CB,$$

or
$$\mu = \frac{AC}{CB} = \tan \alpha.$$

Therefore a body rests in limiting equilibrium on the

inclined plane, when the angle of friction is equal to the inclination of the plane; and if the inclination of the plane is adjustable, the angle of friction can be found by tilting the plane till motion is about to take place.

A better way of finding the coefficient of friction is the following:—

C is a horizontal plane forming the upper surface of one body; the other body A, of mass N, rests on this by a plane face. A horizontal string passes from A over a pulley E, mounted on friction wheels, and from the free end of the string is hung a box in which masses can be placed. These masses are gradually increased till A begins to move. If M be the suspended mass when motion begins, Mg is the horizontal stress between A and C, and $\frac{M}{N}$ is the coefficient of friction.

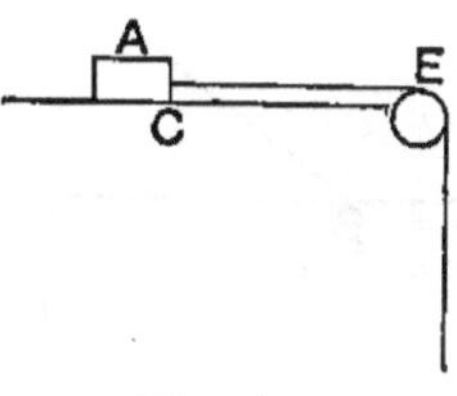

Fig. 89.

§ 14. Initial Motion under Friction.

The results just obtained apply only to particles. The equilibrium of an extended body resting on a rough surface may be broken in two ways—the body may slip along the surface or roll on some point in its base.

If the surface is rough enough, the body will roll before it slides; an example will make this clearer.

A prism, whose section by a plane perpendicular to its edges is a right-angled triangle ABC, rests on a horizontal plane (*Fig.* 90). *To the middle point of the upper edge a horizontal cord is attached, which passes over a smooth pulley and carries a weight W. To find the initial motion of the prism if W is gradually increased.*

If there is equilibrium, the forces acting on the prism are

its weight G, the horizontal tension W, and the resistance of the plane BC which has components, vertical and horizontal, equal to G and W.

Therefore sliding begins if $W > \mu G$.

But the equilibrium may be broken by motion round C.

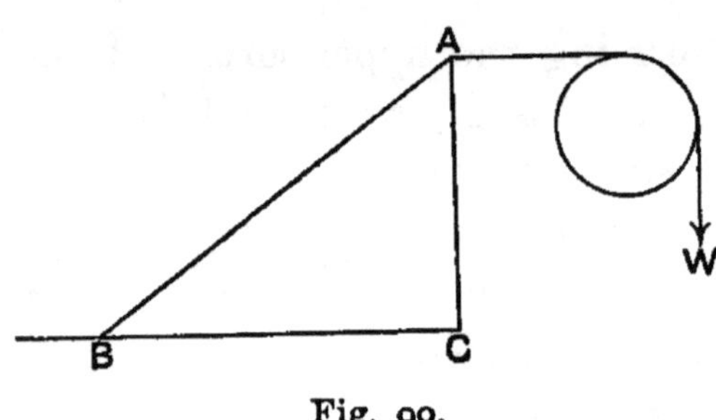

Fig. 90.

Let the resultant upward force exerted by the plane on the prism act through a point in BC, distant x from C, and let

$$AC = b, \ BC = a.$$

Taking moments round C we have as the condition of equilibrium

$$bW + xG = \tfrac{1}{3} aG.$$

Therefore as W increases x diminishes, and when the total resistance acts at C,

$$W = \frac{a}{3b} G.$$

For any greater value of W there cannot be equilibrium; the prism begins to turn on its edge through C.

Therefore, as W increases, equilibrium is broken by rolling if $\mu > \frac{a}{3b}$, and by sliding if $\mu < \frac{a}{3b}$.

Friction on the Inclined Plane.

A particle is on the point of sliding up a rough inclined plane under a force P making an angle α *with the line of greatest upward slope. To find the condition of equilibrium.*

Let W be the weight of the particle; R, μR the normal and tangential pressures of the plane on the particle, θ the inclination of the plane to the horizon.

Resolving along the plane,

$$P \cos a - W \sin \theta - \mu R = 0.$$

And resolving at right angles to the plane,

$$P \sin a - W \cos \theta + R = 0.$$

Hence $P (\cos a + \mu \sin a) = W (\sin \theta + \mu \cos \theta)$.

This is also the condition for motion with uniform velocity, provided that the coefficient of friction *under motion* is substituted for μ.

If the particle is on the point of sliding down the plane,

$$P (\cos a - \mu \sin a) = W (\sin \theta - \mu \cos \theta).$$

Friction on the Screw.

Let a be the pitch and b the radius of the screw, and let the driving force P be on the point of overcoming the resistance W. At each point H where the thread is in contact with the sister screw, it is subject to a force r perpendicular to its surface, and to a force μr tangential to its surface, opposing the motion, and perpendicular to the plane through H and the axis of the screw.

Let R be the sum of all the resistances r at the point of contact of the screws.

Resolving parallel to the axis of the screw,

$$W - R \cos a + \mu R \sin a = 0. \qquad (1)$$

To obtain another relation, suppose that the screw is turned through one revolution, and let s be the displacement of a point on the thread measured along the path by which the point travels.

Then $s \cos a = 2\pi b$, if the thread is very narrow;

$s \sin a = h$, where $\frac{1}{h}$ is the number of threads per unit length.

If a is the radius of the arm at the end of which P is applied, the work done in the displacement is

$$2\pi a \,.\, P - Wh - \mu R s.$$

Equating this to zero, and substituting for s and h, we have

$$aP\cos\alpha = b\,(W\sin\alpha + \mu R). \qquad (2)$$

Substituting in (2) the value of R obtained from (1),

$$(\cos\alpha - \mu\sin\alpha)\,aP\cos\alpha = (\cos\alpha - \mu\sin\alpha)\,bW\sin\alpha + b\mu W$$
$$= bW\cos\alpha\,(\sin\alpha + \mu\cos\alpha);$$

$$\text{or}\quad P = \frac{bW}{a} \cdot \frac{\sin\alpha + \mu\cos\alpha}{\cos\alpha - \mu\sin\alpha}.$$

Action at a Rough Hinge.

It has been shown that if two smooth coaxial cylinders press on each other, the pressures reduce to forces which all meet the axis of the cylinders and can be replaced by a single force. If the hinge is rough, the pressures are not necessarily directed towards the axis, and the action does not necessarily reduce to a single force.

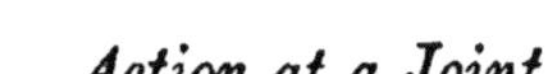

Action at a Joint.

An illustration will best explain this.

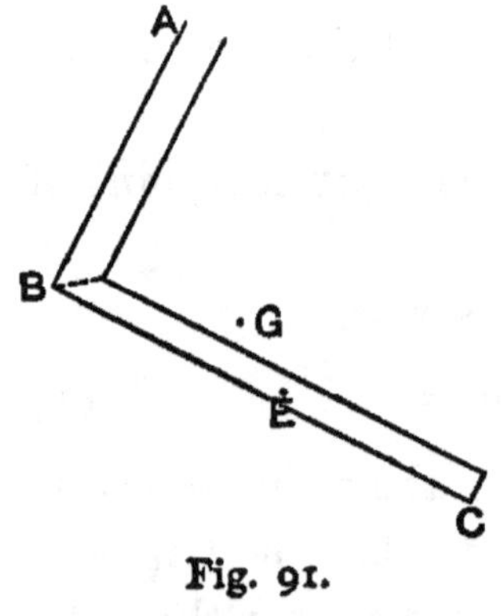

Fig. 91.

ABC is a bar, jointed at B, and suspended by a smooth hinge at A. When in equilibrium it sets itself so that G, the centroid, is vertically below A.

Let E be the centroid of BC.

BC is in equilibrium under its own weight, equivalent to a force W at E, and the action of the joint.

Hence the action of this joint must be equivalent to an

upward force W at B and a couple of moment $W.BE\cos\theta$, where θ is the inclination of BC to the horizon.

The complexity of the force at the joint is due to one part being under a pressure and the other under tension. With reference to this, consult chapter VII, § 13.

Problems on the equilibrium of systems of bars can be very elegantly treated by the methods of Graphical Statics, but space does not permit us to discuss them here.

Examples.

When the forces are all in one plane it is necessary to express the following conditions:—

(1) The sum of the component forces in any direction vanishes; this condition is satisfied when the sum of the components vanishes for each of two perpendicular directions.

(2) The sum of the moments of the forces round any point vanishes.

These may be replaced by the following conditions:—

The sum of the moments of the forces round any three non-collinear points in the plane vanishes.

For let A, B, C be the three points, F, G, H, etc. the forces. The moment of F round A is equal to the moment of F round B together with the moment round A of a force F at B.

Applying this result to each force we have

$$M_A = M_B + B_A$$

when M_A, M_B are the sums of the moments of the forces round A and B respectively, and B_A is the moment round A of the resultant of F, G, H,... supposed to act at B.

Similarly $$M_C = M_B + B_C.$$

Now if M_A, M_B, M_C are all zero, $B_A = 0$ and $B_C = 0$. But B is not on the straight line AC, and therefore the resultant of F, G, H... at B cannot pass through both A and C. Hence, as its moment round both points is zero, it must vanish, and the conditions (1) and (2) are satisfied.

1. A circular cylinder rests with its horizontal axis on a rough inclined plane, and is supported by a string coiled round its middle section and supporting a weight. To find the condition of equilibrium.

Let W be the weight of the cylinder, P the weight suspended from the cord. Since these forces are vertical, the resistance at E must be vertical also. Let α be the inclination of the plane.

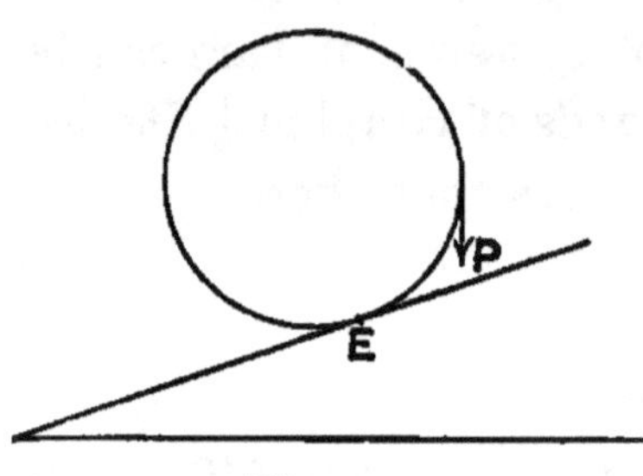

Fig. 92.

Taking moments round E

$$W \sin \alpha = P(1 - \sin \alpha).$$

This relation determines the weight P.

2. A perfectly flexible uniform string hangs from points A and B. To find the conditions of equilibrium.

The tension at P reduces to a single force directed along the tangent at P.

Let T_1, T_2 be the tensions at P and Q, θ_1 and θ_2 the angles which the tangents at P and Q make with the horizon.

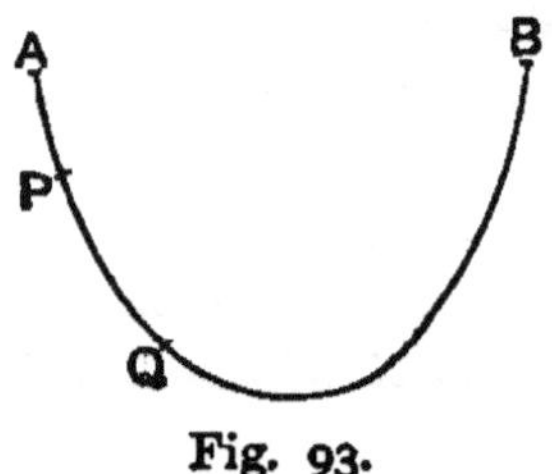

Fig. 93.

The string between P and Q is at rest under its own weight and the tensions; let the arcs AP, AQ be equal to s_1 and s_2 respectively, and let ρ be the mass of unit length of the string.

Then $$T_1 \cos \theta_1 = T_2 \cos \theta_2,$$

$$T_1 \sin \theta_1 - T_2 \sin \theta_2 = \rho (s_2 - s_1) g.$$

Hence the horizontal component of the tension is everywhere the same and is equal to t, the tension at the lowest point.

Therefore $$T_1 \sin \theta_1 - T_2 \sin \theta_2 = t(\tan \theta_1 - \tan \theta_2).$$

And $$\rho g s_1 + t \tan \theta_1 = \rho g s_2 + t \tan \theta_2.$$

The curve is therefore such that $s + \dfrac{t}{\rho g} \tan \theta =$ constant, where s is measured from A.

Also the centre of gravity of the arc PQ lies vertically above the intersection of the tangents at P and Q, and since the equilibrium is stable it lies lower than the centre of gravity of any other arc of equal length joining P and Q.

3. AB and CD are equal rods that can turn freely round a rivet O passing through their middle points; the ends A and C are joined by a thread of given length l, the plane of the rods is vertical, and the system rests with A and C on a horizontal plane. Given weights P and Q are hung at D and B; find the tension of the thread and the stress on the rivet, the weights of the rods being neglected.

Let X and Y be the forces exerted by the plane at A and C. The system formed by the two rods is at rest under the forces X, Y, P, and Q.

Resolving them vertically, $X+Y=P+Q$.

Taking moments round O, $\frac{l}{2}(X-P)=\frac{l}{2}(Y-Q)$, whence

$$X-Y=P-Q.$$

Therefore $$X=P, \quad Y=Q.$$

Let T be the tension in the string.

The rod AB is at rest under the forces X (or P), Q, T and the force exerted on it by B at the rivet. If the components of this latter force are x and y, we have by resolving the forces

$$x+T=0,$$

$$y+P-Q=0.$$

Taking moments round A, we have, if $AD=a$,

$$\frac{a}{2}x+\frac{l}{2}y=lQ.$$

We have then three equations from which x, y, and T can be found.

4. A table of weight W is supported on three legs, which meet the horizontal plane through the centre of mass in A, B, C. The distances of the centre of mass from BC, CA and AB being p, q, r, it is required to find the pressure on each leg.

Let a, b, c denote the sides BC, CA, AB of the triangle ABC; A, B, C the angles opposite to a, b, c.

Let R, S, T be the pressures on A, B, C.

Taking moments round the axis BC we have

$$R \cdot c\sin B = p \cdot W.$$

Similarly by taking moments round CA and AB,

$$S \cdot a \sin C = q \cdot W,$$

$$T \cdot b \sin A = r \cdot W.$$

Hence R, S and T are known.

Since $R+S+T=W$, we have

$$ap+bq+cr = 2\Delta ABC.$$

Equilibrium is impossible if the centre of mass of the table lies without the triangle.

5. A cube is placed with one edge on a rough horizontal plane with coefficient of friction μ, and a parallel edge on a smooth plane inclined at an angle 45° to the vertical; if θ is the inclination of the base of the cube to the horizon when on the point of sliding down, show that

$$(1+3\mu)\tan\theta = 1-\mu.$$

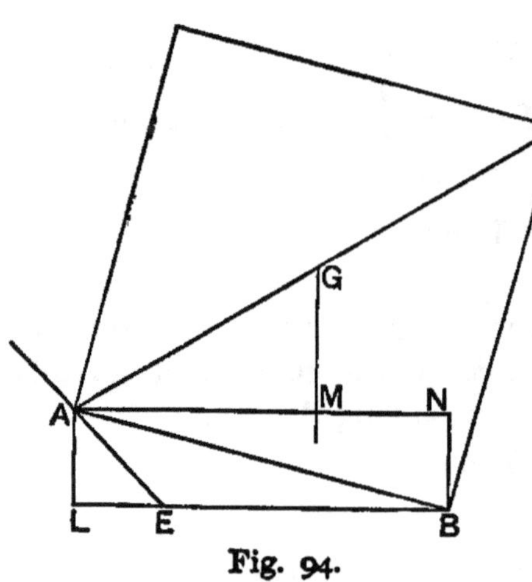

Fig. 94.

Let the plane of the figure be a vertical plane through the centre of mass of the cube, perpendicular to the inclined plane.

AB is the section of the lowest face of the cube, G the centroid of the cube, and if $AB = 2a$, $AG = a\sqrt{2}$.

The resistances at the edges of the cube reduce to a single force R at A perpendicular to AE, a vertical force S at B, and a horizontal force μS at B parallel to BE. The weight of the cube is W.

Draw the vertical lines AL, BN, GM, and the horizontal line AMN through A.

Then

$$AM = AG\cos(45-\theta) = a\sqrt{2}\cos(45-\theta) = a(\sin\theta+\cos\theta),$$

$$AN = 2a\cos\theta.$$

Resolving R into its vertical and horizontal components, and equating the vertical and horizontal forces separately to zero,

$$W-\frac{R}{\sqrt{2}}-S=0, \quad \frac{R}{\sqrt{2}}-\mu S=0.$$

Therefore $$W=(\mu+1)S. \qquad (1)$$

Also the sum of the moments round A vanishes.

Therefore $$AM.W-AN.S+AL.\mu S=0,$$

or $$aW(\sin\theta+\cos\theta)-2a\cos\theta.S+2a\sin\theta.\mu S=0.$$

Substituting from (1) and dividing by aS,

$$(3\mu+1)\sin\theta+(\mu-1)\cos\theta=0.$$

Which is the relation required.

6. A horizontal bar of length 10 feet is suspended by vertical strings attached to its extremities. If the mass of the bar is 10 lbs. and a mass of 4 lbs. is attached to it at a distance from 4 ft. from one end, find the tensions in the strings.

7. A ladder of length 30 feet rests in limiting equilibrium on rough ground, with its upper end against a smooth vertical wall 24 feet above the ground. If the weight of the ladder is 80 lbs., find the pressure on the ground and the coefficient of friction.

8. A mass of 6 lbs. is supported on a smooth inclined plane, rising 5 feet in 18, by means of a string inclined at an angle 30° to the plane. Find the tension in the string.

9. Two masses P and W balance on the Wheel and Axle. Show that if they are interchanged, W will after one second be descending with velocity $\frac{gW(W-P)}{W^2-WP+P^2}$.

10. A mass of 12 lbs. rests on a rough plane inclined 30° to the horizon. What force must be applied to it at an angle 30° to the vertical in order that it may be on the point of moving up the plane, the coefficient of friction being $\frac{1}{2}$?

11. A barrel weighing 5 cwt. is lowered into a cellar down a smooth slide inclined at 45° to the vertical. It is lowered by two ropes passing under the barrel, one end of each rope being fixed, while two men pay out the other ends of the ropes. What pull in

pounds weight must each man exert that the barrel may descend with uniform velocity?

12. In example 2, if the length of the string is l and a is the angle which each upper end makes with the horizon, $t \tan a = \frac{1}{2} g \rho l$.

13. A chain hangs between two points. Prove that the square of the tension at any point P is equal to the sum of the squares of the tension at the lowest point and of the weight of chain between P and the lowest point.

14. A right cone, diameter of base a, height b, is placed on a plane with coefficient of friction μ. Find the greatest value μ may have that the cone may slide without turning over when the inclination of the plane is gradually increased.

15. A uniform rod rests with one end on a rough horizontal plane and the other end on a smooth plane inclined at 30° to the horizon. Prove that if the rod is on the point of sliding down, its inclination θ to the horizon is given by

$$2 \mu \tan \theta = 1 - \mu \sqrt{3},$$

μ being the coefficient of friction.

16. Prove that any force can be reduced to another force and a couple in the same plane with it; and hence that any system of coplanar forces may be reduced to a single force acting at a specified point together with a couple.

17. A particle of mass 50 lbs. hangs from two strings, respectively 7 and 8 feet long, the upper ends of which are fastened to pegs 9 feet apart in the same horizontal line. Find the tension in each string.

18. An uniform beam 12 feet long rests with one end against the base of a wall 20 feet high. The beam is supported by a rope 13 feet long attached to the top of the beam and the summit of the wall. Find the tension in the rope, neglecting its weight and assuming the mass of the beam to be 100 lbs.

19. A beam AB of length l leans against a rough vertical wall, the end A being prevented from sliding along the smooth horizontal plane AD by a string AD attached to the wall. Find the tension in the string, if μ is the coefficient of friction and a the inclination of the beam to the vertical.

20. A step-ladder in the form of the letter A, with both legs inclined at an angle α to the vertical, is placed on a horizontal floor and is held up by a cord connecting the middle points of its legs, there being no friction anywhere; prove that when a weight W is placed on the top of the ladder the tension in the cord is increased by $W \tan \alpha$.

Also if W is placed on a step at a height from the floor equal to $\frac{1}{n}$ of the height of the ladder, the increase of tension is $\frac{1}{n} W \tan \alpha$.

21. A uniform triangular plate with one vertex A resting on a smooth horizontal plane is supported in a vertical plane by a cord attached to the vertices B and C, and passing over a fixed pulley at D. If BC is horizontal, show that the pressure on the plane is one-third of the weight of the plate.

22. A cylinder of radius r whose axis is fixed horizontally touches a vertical wall along a generating line. A flat beam of uniform material of length $2l$ and weight W rests with its extremities in contact with the wall and cylinder, making an angle 45° with the vertical.

Prove that in the absence of friction

$$\frac{l}{r} = \frac{\sqrt{5}-1}{\sqrt{10}},$$

that the pressure on the wall is $\frac{1}{2}W$, and on the cylinder is

$$\frac{1}{2\sqrt{5}} W.$$

CHAPTER VI.

GRAVITATION.

§ 1. THE Law of gravitation asserts that any two particles of matter attract one another with a force which is proportional directly to the product of their masses, and inversely to the square of their distance apart.

The evidence in favour of this law is mainly derived from the consideration of the elliptic orbits described by the planets round the Sun.

A few properties of the ellipse that will be useful are proved in §§ 2–4.

§ 2. An ellipse may be most simply obtained in the following way. Let ABA' (Fig. 95) be a circle with centre C. From a point Q on the circle draw an ordinate QN perpendicular to a diameter ACA', and on QN take a point P such that $PN : QN$ is a ratio < 1.

If different positions of Q on the circle are taken, and the fraction $\frac{PN}{QN}$ is the same for all, the point P lies on an ellipse.

Since the point Q may lie above or below the line ACA', the ellipse consists of two precisely similar portions, one above, the other below ACA'.

C is called the centre of the ellipse, any chord through C

is called a diameter, and ACA', being obviously the longest diameter, is called the major axis.

If BCB' be the diameter of the circle perpendicular to

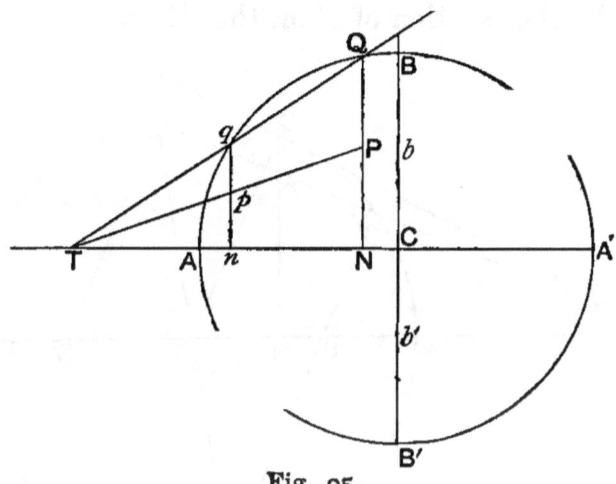

Fig. 95.

ACA', b, b' the points where it meets the ellipse, bCb' is called the minor axis of the ellipse. It divides the ellipse into two similar parts.

§ 3. Let QN, qn be two ordinates of the circle ABA', P and p corresponding points on the ellipse.

If T be the point in which Qq meets AA',

$$\frac{TN}{Tn} = \frac{QN}{qn} = \frac{PN}{pn}, \text{ by the definition of the ellipse.}$$

Therefore Pp passes through T.

Now suppose that q moves along the circle till it coincides with Q; then p moves along the ellipse till it coincides with P, and the chords Qq, Pp approach and ultimately coincide with the tangents at Q and P to the circle and ellipse respectively.

Hence if QN, QT be the ordinate and tangent at Q (Fig. 96), TP is the tangent at P to the ellipse through P whose major axis is AA'.

Let TP meet the circle in Y and Y'; and draw YS, $Y'S'$ perpendicular to TP, meeting ACA' in S and S'.

We shall show that S and S' are fixed points in ACA' whatever be the position of P on the ellipse.

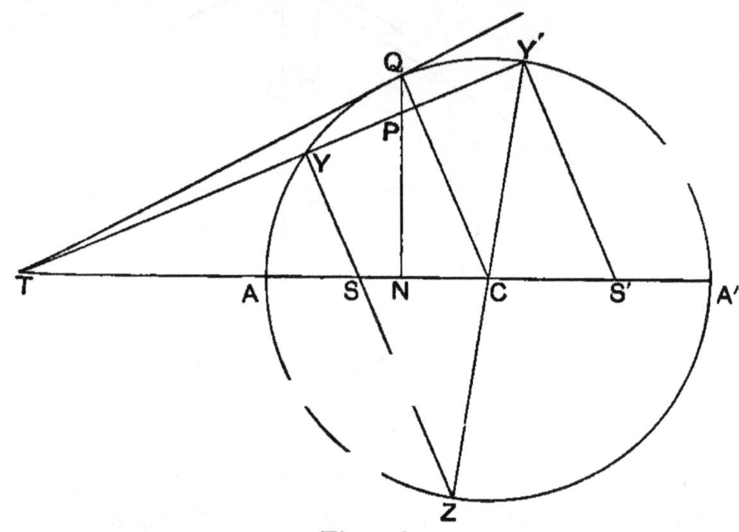

Fig. 96.

Produce YS to meet the circle in Z, and join CQ.

Then, since ZYY' is a right angle, ZY' passes through C (Euclid III. 31).

In the triangles CSZ, $CS'Y'$, the angles CSZ, SCZ are equal to $CS'Y'$, $S'CY'$, each to each, and $CY' = CZ$.

Therefore (Euclid I. 26) $CS = CS'$ and $SZ = S'Y'$.

Now the triangles PNT, SYT are similar.

Therefore (Euclid VI. 4) $\dfrac{PN}{NT} = \dfrac{SY}{YT}$.

Similarly the triangles PNT, $S'Y'T$ are similar, and

$$\frac{PN}{NT} = \frac{S'Y'}{Y'T}.$$

Therefore $$\frac{PN^2}{NT^2} = \frac{SY.S'Y'}{TY.TY'} = \frac{SY.SZ}{TY.TY'}.$$

But $TY.TY' = TQ^2$.

Therefore $\frac{PN^2}{NT^2} = \frac{SY.SZ}{TQ^2}$ or $\frac{PN^2}{SY.SZ} = \frac{NT^2}{TQ^2}$.

Again, in the similar triangles CQT, QNT,

$$\frac{NT}{TQ} = \frac{QN}{CQ} = \frac{QN}{CA} = \frac{PN}{Cb}.$$

Therefore $$\frac{PN^2}{Cb^2} = \frac{NT^2}{TQ^2} = \frac{PN^2}{SY.SZ}.$$

And $SY.SZ = Cb^2$.

Now $SY.SZ = AS.SA'$ (Euclid III. 35)

$= CA^2 - CS^2$.

Therefore $CA^2 - CS^2 = Cb^2$,

or $CS^2 = CA^2 - Cb^2$.

Therefore S is a fixed point, and so is S' for $CS = CS'$. S and S' are called the foci of the ellipse.

§ 4. The following property of the circle is useful.

If QN be the ordinate of a point Q, QT the tangent at Q, meeting ACA' in T, TYY' any chord through T, meeting the ordinate QN in P.

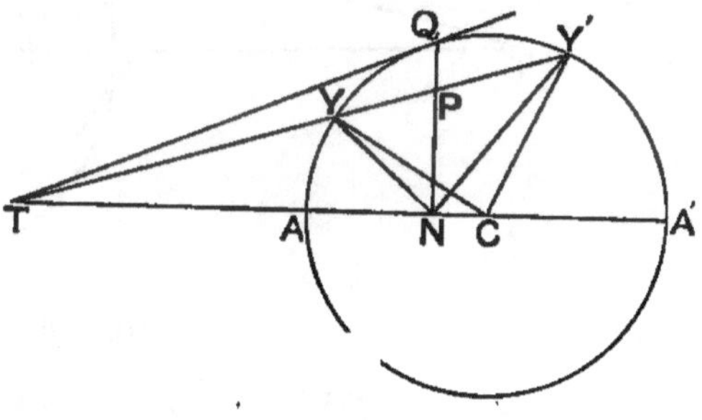

Fig. 97.

$$\frac{TY}{TY'} = \frac{PY}{PY'}.$$

For $CN : CQ :: CQ : CT$, since CQN, CTQ are similar triangles.

And $CQ = CY = CY'$.

Therefore $CN : CY :: CY : CT$.

And the triangles *CNY*, *CYT* are similar (Euclid VI. 6).

Therefore $\angle CNY = \angle CYT.$

And $\angle YNT = \angle CYY'.$

In the same way it can be shown that *CNY'*, *CY'T* are similar triangles.

And that $\angle Y'NC = \angle CY'Y.$

But the angles *CYY'*, *CY'Y* are equal.

Therefore *NT* bisects the angle *YNY'* externally, and *NQ*, being at right angles to *NT*, bisects *YNY'* internally.

Therefore (Euclid VI. 3 and A) $\dfrac{PY}{PY'} = \dfrac{NY}{NY'} = \dfrac{TY}{TY'}.$

Let *TP* be the tangent at a point *P* on the ellipse.

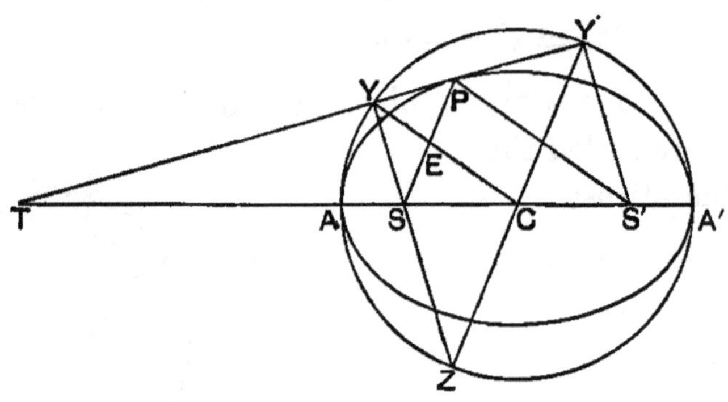

Fig. 98.

Then $$\frac{PY}{PY'} = \frac{TY}{TY'} = \frac{SY}{S'Y'} = \frac{SY}{SZ}.$$

Therefore (Euclid VI. 2) *SP* is parallel to *Y'CZ*.

Similarly *S'P* is parallel to *CY*.

Therefore $\angle SPY = \angle CY'Y = \angle CYY' = \angle S'PY'.$

And *SYP*, *S'Y'P* are right angles.

Therefore the triangles *SPY*, *S'PY'* are similar,

$$\text{and} \quad \frac{SP}{SY} = \frac{S'P}{S'Y'}, \qquad (1)$$

a result which we shall require later.

Again, let SP and CY intersect in E.

Because $SC = CS'$ and EC, $S'P$ are parallel,

$$EC = \tfrac{1}{2} S'P, \text{ and } PE = \tfrac{1}{2} SP.$$

Therefore $SP + S'P = 2\,(PE + EC)$.

But $PE = EY$, for the angles EYP, EPY are equal.

Therefore $SP + S'P = 2\,(CE + EY) = AA'$. (2)

The area of the ellipse is to the area of the circle as PN is to QN or as Cb is to AC.

Therefore the area of the ellipse is $\pi AC^2 \dfrac{Cb}{AC}$, or $\pi\, AC \,.\, Cb$.

§ 5. Let us now consider a point P moving on a curved path so that the radius vector SP, drawn to P from a fixed point S, sweeps over equal areas in equal times.

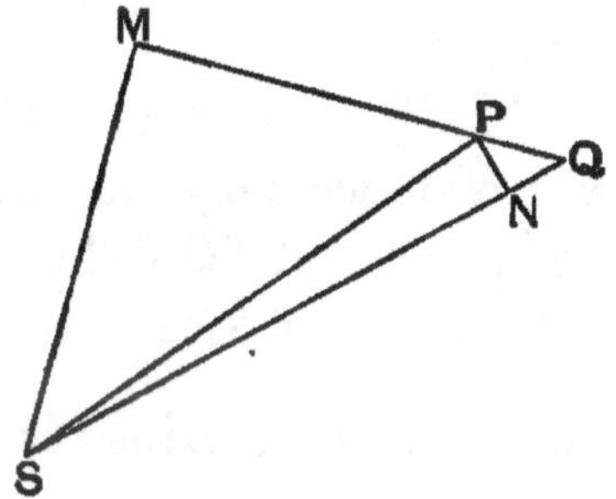

Fig. 99.

Let P, Q be positions of the moving point at the beginning and end of a time t.

Join PQ, and draw SM perpendicular to PQ.

When t is made indefinitely small, the chord PQ becomes the tangent at P to the path.

Denote the length SM by p, and let v be the velocity at P.

Since the area $SPQ = \frac{1}{2}\, SM \,.\, PQ$, the area described in unit time is

$$\tfrac{1}{2} SM \,.\, \frac{PQ}{t} \text{ or } \tfrac{1}{2} pv.$$

Thus $vp = h$, where h is a constant for all points of the path.

Hence the moment of the velocity at P round S remains the same throughout the motion.

Now the moment of a velocity, like that of a force, is equal to the sum of the moments of its components, and the velocity at Q is compounded of the velocity at P, and the velocity gained in passing from P to Q.

Therefore since the moments of the velocities at Q and at P are equal, the moment of the velocity gained between P and Q is zero. That is, either no velocity is acquired, or the acceleration is directed towards S.

But there must be acceleration, as the path is curved, and therefore the acceleration of the moving point is directed towards S.

The result, $vp = h$, may be expressed in another way.

For if PN be perpendicular to SQ, the area SPQ is $\frac{1}{2} PN \,.\, SQ$.

If PSQ be a very small angle whose circular measure is θ, PN is sensibly an arc of a circle with centre S and radius SP, and SQ differs insensibly from SP.

Therefore if $SP = r$, the area of the triangle SPQ is $\frac{1}{2} r^2 \theta$, and the area described in unit time is $\frac{1}{2} r^2 \frac{\theta}{t}$, or $\frac{1}{2} r^2 \omega$, where ω is the angular velocity round S.

Therefore $r^2 \omega = h$, a constant.

§ 6. Let the path of the moving point be an ellipse, the radius vector from one focus S describing equal areas in equal times.

Denote the lengths of CA and Cb by a and b.

The velocity at P is $\frac{h}{SY}$ or $\frac{h}{b^2} S'Y'$, since $SY \,.\, S'Y' = b^2$.

Hence if unit velocity is represented by a line of length $\frac{b^2}{h}$, the hodograph is obtained by turning the circle ABA' through a right angle round S', and the acceleration at P is parallel to $Y'C$, that is, it is along PS.

Again, since CY' is parallel to SP, the angular velocity of Y' round C is ω, and the acceleration at P is

$$\frac{ha\omega}{b^2}, \text{ or } (\S\ 5)\ \frac{h^2 a}{b^2 r^2}.$$

§ 7. Kepler's Laws of Planetary Motion.

The planets describe orbits round the sun in accordance with the following laws, which were first enunciated by Kepler in 1609.

Law I. The areas swept out by a line drawn from the sun to a planet are proportional to the times taken in describing them.

Law II. The orbit of a planet is an ellipse with the sun in one of the foci.

Law III. The squares of the periodic times of different planets are proportional to the cubes of the major axes of their orbits.

Since the sun and planets are extended bodies, it must be understood that the line mentioned in Law I is that which joins the centres of the sun and planets, and that the centre of the sun is the focus of the ellipse described by the centre of the planet. But as the dimensions of the planets are very small compared with their distances from the sun, we shall treat each planet as a particle moving along the path of the centre of the planet.

The first law shows that the acceleration of a planet is towards the sun, and the second law shows that the acceleration is inversely proportional to the square of the distance of the planet from the sun.

This statement is true for each planet, but we have no means of comparing the action of the sun on different planets till we interpret the third law.

If $\frac{1}{2}h$ be the area described round the sun in unit time, the area of the ellipse πab is described in time $\frac{2\pi ab}{h}$.

This is called the periodic time of the planet which describes the ellipse, and is denoted by T.

Hence $$\frac{a^3}{T^2} = \frac{h^2 a}{4\pi^2 b^2}.$$

Now $\frac{a^3}{T^2}$ is the same for all orbits described round the sun, by Kepler's third law. Therefore $\frac{h^2 a}{b^2}$ is the same for all the planets.

Hence the accelerations of different planets are inversely proportional to the squares of their distances from the sun.

§ 8. The Law of Gravitation.

If a planet describes a circular orbit with uniform velocity its centripetal acceleration can be found, but no inference can be drawn as to the dependence of the acceleration on the distance between the planet and the body round which it revolves, for the distance always remains the same.

But if several planets describe circular orbits uniformly round a common centre, and the squares of the periodic times are as the cubes of the radii of the orbits, it is clear that the accelerations of the several planets are inversely proportional to the squares of their distances from the centre.

For in a circle of radius A described with velocity v the acceleration a is $\frac{v^2}{A}$, but $v = \frac{2\pi A}{T}$.

Therefore $$a = \frac{4\pi^2 A}{T^2} = \frac{4\pi}{A^2} \cdot \frac{A^3}{T^2}.$$

But $\frac{A^3}{T^2}$ is constant for all orbits round the common centre.

Therefore $$a \propto \frac{1}{A^2}.$$

This leads us to consider some new phenomena.

Almost at the same time that Kepler announced his laws of planetary motion, Galileo by his application of the telescope to astronomy was enabled to detect the four satellites of Jupiter. These bodies describe approximately circular orbits round Jupiter, they form on a small scale a system resembling the planetary system, and it will be seen from the following table that they satisfy Kepler's third law very closely.

Satellite.	T hrs.	T min.	A *	$\frac{A^3}{T^2}$ †
1	42	27·6	6·04853	$10^{-5} \times$ 3·4096
2	85	14·6	9·62347	3·4070
3	171	42·6	15·35024	3·4076
4	400	31·8	26·99835	3·4075

It is probable then that the influence of Jupiter on these satellites is of the same kind as that which the sun exerts on the planets.

Saturn has eight satellites, and Uranus five, whose motions round their respective primaries obey the same law.

Our conception of force as stress leads us to regard these motions as due to the action of the body round which they take place; thus we say that the sun attracts the planets,

* The equatorial radius of Jupiter is taken as 1.

† T having been reduced to minutes.

and the planets their satellites, and both these attractions are as the inverse square of the distance.

Now it appears (§ 7) that if two planets, e. g. Mars and Venus, could be brought to equal distances from the sun, they would be equally accelerated towards his centre. But it is unlikely that the masses of all the planets are equal, considering their known inequalities in size.

Therefore the force exerted by the sun on a planet is proportional to the mass of the planet, other things being equal.

Similarly the force exerted by Jupiter on a satellite is proportional to the mass of the satellite, if the law of the inverse square is true; and the force exerted by Jupiter on the sun (equal and opposite to the force exerted by the sun on Jupiter) is proportional to the mass of the sun.

Therefore if M is the mass of the sun, m the mass of Jupiter or any other planet at a distance r from the sun, the attraction between the sun and planet is proportional to $\frac{Mm}{r^2}$ and may be denoted by $\frac{kMm}{r^2}$, where k does not depend on either the masses or distances of the bodies.

There is at present no evidence that k depends on the physical condition of the attracting masses, or on the medium between them.

If, as is generally done, we regard k as constant, we may advantageously employ a new unit of mass, called the astronomical unit and defined as follows:—

The mass of each of two equal particles which impart to one another an acceleration of one centimetre per-second per-second when placed 1 centimetre apart is the astronomical unit of mass, in the centimetre-second system of units.

When this unit is used, k becomes equal to 1 and $\frac{Mm}{r^2}$ is the attraction between two masses m and M at a distance r apart.

Since the planets attract their satellites, they must attract other planets and cause them to deviate from their calculated paths. Such deviations exist, proving the mutual attraction of the planets, but they cannot be discussed in an elementary work.

It may, however, be stated that the inconsistency of the calculated and observed motions of the planet Uranus led M. Leverrier and Professor Adams independently to calculate, from the law of gravitation, the orbit of an unknown planet which was presumed to be the cause of the disturbance. Both astronomers predicted the magnitude, orbit, and position of this planet with considerable accuracy; and guided by their instructions observers had little difficulty in discovering Neptune, which is the disturbing body.

The discovery of this planet was a striking confirmation of the theory of gravitation.

§ 9. Correction of Kepler's Third Law.

The sun, the planets, and their satellites form a system of bodies which move freely under their mutual attraction. The centre of mass of this system can be regarded as fixed, and since the mass of the sun is very large, compared with that of the planets, the centre of mass is never very far from the sun's centre.

Let G be the centre of mass of the sun S and a planet P. The motion of G is only affected by the action of other planets, and, neglecting this action, we may regard G as at rest.

If s and p are the masses of the sun and planet,

$$\frac{SG}{SP}=\frac{p}{s+p}.$$

And if c is the acceleration of P relatively to S, $\frac{cp}{s+p}$ is the acceleration of G relatively to S, and $c-\frac{cp}{s+p}$ or $\frac{cs}{s+p}$ is the acceleration of P relatively to G.

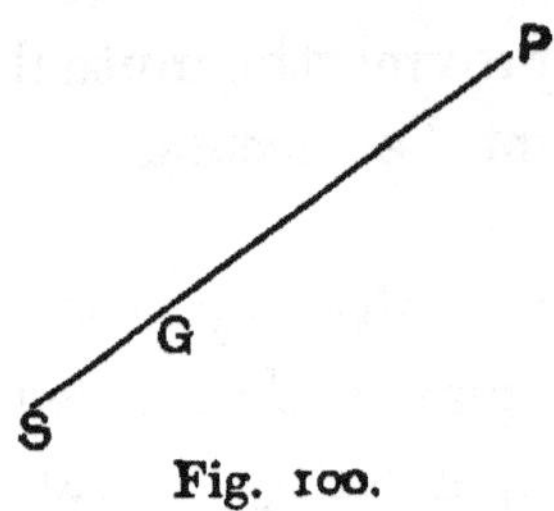

Fig. 100.

Now $2a$, $2b$ being the major and minor axis of the orbit round the sun, and T the periodic time, it has been shown that the acceleration relatively to the sun is $\frac{ah^2}{b^2r^2}$, where $h=\frac{2\pi ab}{T}$.

Therefore the acceleration relatively to the sun is $\frac{4\pi^2a^3}{T^2r^2}$,

and the acceleration relatively to G is $\frac{4\pi^2a^3s}{(s+p)T^2r^2}$.

Since this is the true acceleration the force exerted by the sun on the planet is $\frac{4\pi^2spa^3}{(s+p)T^2r^2}$.

But by the law of gravitation it is $\frac{sp}{r^2}$.

Therefore $\frac{a^3}{T^2(s+p)}=\frac{1}{4\pi^2}$, a constant.

Therefore Kepler's Third Law requires amendment and must run as follows.

The cubes of the major axes of the orbits are proportional to the squares of the periodic times and the sum of the masses of the sun and planet.

It is found experimentally that Kepler's Third Law is

not quite accurate, and requires the amendment indicated here. The correction is small, and, having indicated its existence, we shall for the future neglect it. It becomes however very important in the investigation of double stars, since if the periodic time and the mean distance of the orbit described by one star about the other can be found, the total mass of the two stars can be determined in astronomical units.

§ **10**. We can now compare the masses of the members of the solar system which possess satellites.

Let a, a' be the mean distances of two satellites from their primaries, T, T' the periodic times of the satellites, s, s' the masses of the primaries.

Then $$\frac{a^3}{T^2} = \frac{s}{4\pi^2}, \quad \frac{a'^3}{T'^2} = \frac{s'}{4\pi^2}.$$

Therefore $$\frac{s}{s'} = \frac{a^3 T'^2}{a'^3 T^2}.$$

Let us compare roughly the masses of the earth and sun. The earth itself is a satellite of the sun, performing its orbit in 365·26 mean solar days, with a mean distance of 93,000,000 miles from the sun. The moon describes its orbit round the earth in 27⅓ days, at a mean distance of 238,000 miles.

It thus appears that the mass of the sun is about 350,000 times that of the earth.

The following table, extracted from Herschel's Astronomy, supplies sufficient data for the comparison of the masses of the principal planets, and for the verification of Kepler's third law.

Satellite.		Radius of Primary.	Mean Distance of Satellite from Primary.[1]	Time of Revolution. days	hrs.	min.
The moon		3950	60·27	27	7	43
Inner Satellite of Mars		2273	2.551	7	39	14
Outer " " "			6.423	30	17	54
Fourth Satellite of Jupiter		45367	27.	16	16	32
Satellites of Saturn	Rhea	38395	9·55	4	12	25
	Titan		22.14	15	22	41
	Hyperion		26·78	21	7	8
	Japetus		64·36	79	7	54
Satellites of Uranus	Ariel	17653	7·4	2	12	29
	Umbriel		10·31	4	3	28
	Titania		16·92	8	16	57
	Oberon		22·56	13	11	7
Satellite of Neptune		20000	11·	5	21	43

The only planets concerning which we have obtained no data are Venus, Mercury, and the Asteroids. The masses of Venus and Mercury have been estimated by determining the disturbing effect of Venus on the Earth's motion, and that of Mercury on the motion of Venus.

The satellites of Mars were only discovered in 1877, and before then the mass of Mars was determined by its effect on the Earth's motion.

The following illustration, given by Sir John Herschel, conveys a good idea of the relative magnitudes and distances of the members of the solar system.

On a level field place a globe two feet in diameter; this will represent the Sun; Mercury will be represented by a grain of mustard seed, moving on the circumference of a circle 164 feet in diameter; Venus by a pea on a circle 284 feet in diameter; the Earth also by a pea on a circle

[1] The distances are given as multiples of the equatorial radius of the Primary.

430 feet across; Mars by a rather large pin's head on a circle 654 feet across; the Asteroids by grains of sand in orbits of from 1000 to 1200 feet; Jupiter by a moderate-sized orange on a circle nearly half-a-mile across; Saturn by a small orange on a circle $\frac{4}{5}$ mile across; Uranus by a full-sized cherry on a circle more than $1\frac{1}{2}$ miles across; Neptune by a good-sized plum on a circle about $2\frac{1}{2}$ miles across.

On the same scale the nearest visible fixed star would be represented by a globe distant 8000 miles from the Sun.

§ 11. Work done by Gravitational Force.

Let us again consider the motion of a particle in an elliptic orbit.

It has been shown in § 6 that if v is the velocity at P (fig. 98),

$$v = \frac{h}{SY} \text{ or } \frac{h}{b^2} S'Y'.$$

Multiplying the two values of v together, and denoting $ah^2 b^{-2}$ by μ,

$$v^2 = \frac{\mu}{a} \cdot \frac{S'Y'}{SY} = \frac{\mu}{a} \cdot \frac{S'P}{SP} \quad [\S\ 4\ (1)].$$

And since $\qquad S'P = 2a - SP \quad [\S\ 4\ (2)],$

$$v^2 = \mu\left(\frac{2}{SP} - \frac{1}{a}\right).$$

If V is the velocity at any other point Q on the ellipse,

$$V^2 = \mu\left(\frac{2}{SQ} - \frac{1}{a}\right),$$

$$\text{and } v^2 - V^2 = 2\mu\left(\frac{1}{SP} - \frac{1}{SQ}\right).$$

Hence, if m is the mass of the moving particle,

$$\tfrac{1}{2} m (v^2 - V^2) = \mu m \left(\frac{1}{SP} - \frac{1}{SQ}\right).$$

Since the term on the left-hand side is the gain of kinetic energy, $\mu m\left(\frac{1}{SP}-\frac{1}{SQ}\right)$ is the work done by the attraction to S in the displacement from Q to P. By § 6 the attraction at P is $\frac{\mu m}{SP^2}$.

Let us choose a very large ellipse, so that Q can be taken at a great distance; then $\frac{1}{SQ}$ is very small, and can be made smaller than any assigned fraction by taking Q far enough off.

If Q is at an infinite distance, $\frac{1}{SQ}=0$, and $\frac{\mu m}{SP}$ is the work done on m in bringing it to P from infinity by any path, since the attraction is a conservative force.

In the following sections masses will be expressed in terms of the astronomical unit.

§ **12.** We have till now regarded the planets and their satellites as moving particles. Observation, however, shows that they are extended bodics of approximately spherical form, but that nevertheless the law of the inverse square of the distance expresses their mutual action with great accuracy.

Assuming then that the law of the inverse square is true for the mutual actions of two particles, we are led to expect that the same law will hold for the mutual action of two spheres, at all events in certain cases. We shall prove this to be true.

The motion of each body in the heavens does not depend on its mass, but only on the masses and positions of neighbouring bodies. The motion of each body is therefore the same as that of a particle of unit mass, subjected to the attraction of the same bodies, and moving with the same initial velocity, as the given body.

It appears from § 11 that if r is the distance of a point P from a fixed particle of mass M, $\frac{M}{r}$ is the work that would be done by the attraction of M in bringing a particle of mass unity from infinity to P

$\frac{M}{r}$ is called the gravitation potential at P due to M.

If there are several particles of mass $M_1, M_2, \ldots M_n$ respectively, at distances $r_1, r_2, \ldots r_n$ from P, the gravitation potential at P is

$$\frac{M_1}{r_1} + \frac{M_2}{r_2} + \ldots + \frac{M_n}{r_n}.$$

If the force is zero in any region the potential there is constant, for the same work is done in passing from infinity to all points of the region.

§ **13. If to all points of a spherical surface equal centripetal forces tend which vary as the inverse square of the distance, a particle within the sphere is in equilibrium under these forces.**

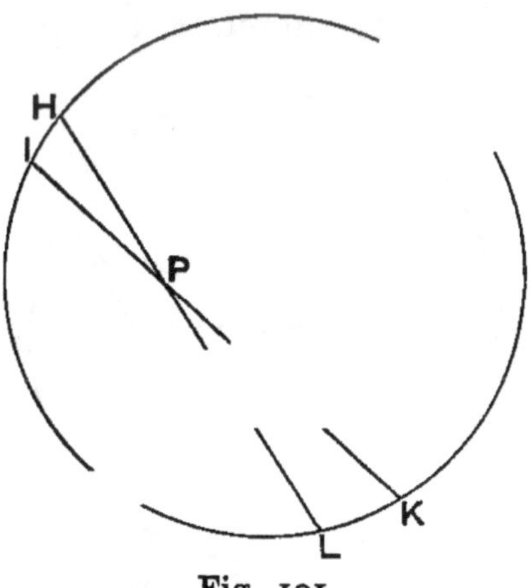

Fig. 101.

Let HKL be the spherical surface, P a point within it.

With P as vertex describe a cone of small aperture cutting the plane of the paper in the lines HPL, IPK.

Then the areas intercepted on the sphere by the cone about IH and LK are to one another in the ratio $IH^2 : LK^2$, or in the ratio $PH^2 : PL^2$.

But if these areas are A, B, the forces due to them at P are in the ratio $\frac{A}{PH^2} : \frac{B}{PL^2}$.

And $\frac{A}{PH^2} = \frac{B}{PL^2}$.

Hence tne forces are equal and their directions are opposite.

Therefore the resultant force at P due to them is zero.

In like manner the whole sphere can be divided into pairs of opposite elements whose actions at P neutralize one another, and therefore the force at P due to the whole sphere is zero.

Hence the potential everywhere within the sphere is the same and has the same value as at the centre.

And if a be the radius of the sphere, M the total mass distributed over its surface in an indefinitely thin layer, the potential at the centre is $\frac{M}{a}$.

This is therefore the potential at any point within the sphere.

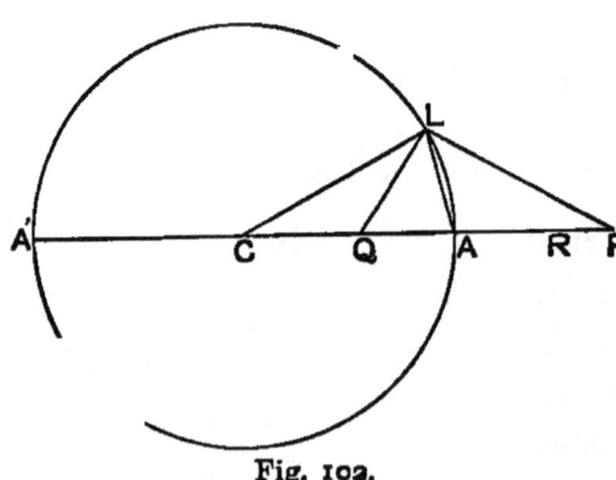

Fig. 102.

§ 14. To find the potential at a point P outside the sphere due to the shell of mass M.

Let C be the centre of the sphere, $A'CA$ the diameter through P, Q a point such that

$$CQ \cdot CP = CA^2,$$

L any point on the sphere.

Join LQ, LA and LP.

Then $CQ : CL :: CL : CP$, and the angle LCQ is common to the triangles CLQ, CPL.

Therefore (Euclid VI, 6) $\angle CLQ = \angle CPL$.

$$\text{But} \quad \angle CLA = \angle CAL = \angle CPL + \angle ALP.$$

$$\text{Therefore} \quad \angle QLA = \angle ALP.$$

Therefore AL bisects the angle QLP,

$$\text{and} \quad QL : LP :: QA : AP.$$

$$\text{But} \quad CQ : CA :: CA : CP.$$

$$\therefore \quad QA : CA :: AP : CP.$$

$$\therefore \quad QL : LP :: CA : CP.$$

Now let u be the potential at Q due to an element of the sphere at L, v the potential at P due to the same element.

$$\text{Then} \quad u : v : \frac{1}{LQ} : \frac{1}{LP},$$

$$\text{or} \quad \frac{v}{u} = \frac{LQ}{LP} = \frac{CA}{CP}.$$

Since $\frac{CA}{CP}$ does not depend on the position of L, the ratio $\frac{v}{u}$ is the same for all elements of the spherical surface.

Therefore if U is the potential at Q, V the potential at P, due to the whole sphere,

$$\frac{V}{U} = \frac{CA}{CP}.$$

And since $U = \frac{M}{CA}$ (§ 13), $V = \frac{M}{CP}$.

Hence the potential of a thin uniform spherical shell of mass M is the same as that of a particle of mass M placed at the centre of the sphere.

The work done by the attraction of the shell in the displacement of a particle m from P to R is

$$Mm\left(\frac{1}{CR}-\frac{1}{CP}\right),$$

and the same work would be done by a particle of mass M at C in the same displacement of m.

Therefore the force exerted by the shell on m is the same as that which a particle of mass M placed at C would exert on m, and if P be the position of a particle m, the force exerted on m by the shell is $\frac{Mm}{CP^2}$.

The substance of the sphere being supposed to lie on its surface, the quantity of matter per unit surface is said to measure the *surface-density.*

If M be the total mass, $\frac{M}{4\pi . CA^2}$ is the surface-density, and is generally denoted by σ.

If the point P lies just outside the surface, CP becomes equal to CA, and the force exerted on a particle, of unit mass, lying just outside the sphere is $\frac{M}{CA^2}$ or $4\pi\sigma$.

This result is of importance in electricity.

A body is said to be uniform when the physical properties of all its parts are precisely similar. For our present purpose its density is the most important of these.

§ 15 (*a*). Attraction of a solid uniform sphere at an external point.

Let M be the mass of the sphere, C its centre. Divide the sphere into indefinitely thin concentric spherical shells each of mass m.

The action of each of these shells on an external particle

is the same as that of a mass m placed at the centre of the sphere. Therefore the action of the whole sphere at an external point is the same as that of a particle of mass M placed at its centre.

This proposition holds also when each thin shell is uniform, whether the sphere is uniform or not.

It is therefore also true when the density of the sphere is not uniform, provided that at points which are equally distant from the centre the density is the same.

COR. The action of the sphere on any external body is the same as the action of a mass M lying at the centre of the sphere.

(*b*) *Force due to a solid uniform sphere at an internal point.*

Let DEF be the sphere, C its centre, P the internal point, at which is a particle of mass unity.

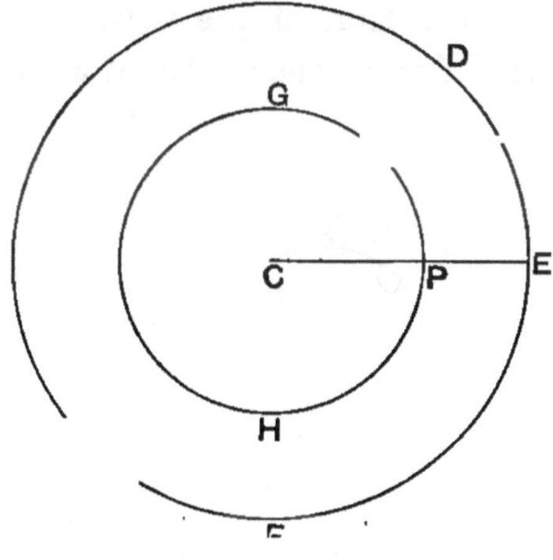

Fig. 103.

With centre C and radius CP describe a sphere PGH.

The solid sphere DEF can be regarded as composed of the sphere PGH and the shell lying between PGH and DEF.

But if the density of the shell is the same at all points equidistant from the centre, it can be divided into thin uniform shells, and the action of each thin shell on an internal point is zero.

Therefore the force at P is due only to the sphere PGH, and is $\frac{N}{CP^2}$, where N is the mass of this sphere.

But $N = \frac{4}{3}\pi\rho \,.\, CP^3$, where ρ is the density of the sphere.

Therefore the force exerted on the particle is $\frac{4}{3}\pi\rho \,.\, CP$.

Or, the force exerted by a solid uniform sphere on an internal particle is an attraction proportional to the distance from the centre.

If the density is not uniform, but is the same at equal distances from the centre, ρ must be taken as the average density of the nucleus PGH.

§ 16. Mutual attraction of Two Spheres.

Let X and Y be two spheres, in each of which the density is either uniform or the same at the same distance from the centre.

Let A and B be the centres of the spheres, M and m their masses, P any particle of the sphere X.

The action of the mass M on Y is unaltered if we suppose M to be concentrated at A, and it is equal and opposite

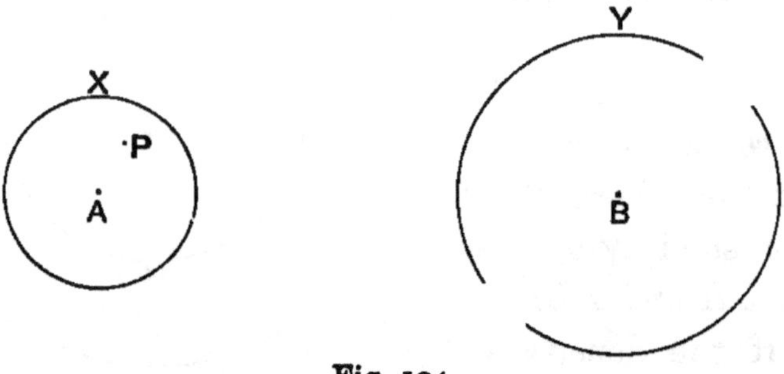

Fig. 104.

to the force exerted by Y on a mass M at A. But this is equal to the force exerted by a mass m at B on a mass M at A, and is therefore $\frac{Mm}{AB^2}$.

We thus see that the same simple law suffices to explain the motions of the planets (and of the satellites round the planets), whether they are large or small, provided that the density of a planet is the same at equal distances from its centre.

The density of a planet is not uniform, but the conditions which determine the density (such as pressure) are probably the same at the same distance from the centre, and we may therefore reasonably regard the planets as satisfying the condition given.

§ 17. Terrestrial Gravitation.

All bodies at a given place on the Earth's surface fall to the ground with the same acceleration g, when the resistance of the air is removed.

We have already explained the motion of the Moon by ascribing it to the Earth's gravitational force.

The mean distance of the Moon from the Earth is about 60 R, where R is the Earth's radius, and therefore, if the Moon were at the Earth's surface, its acceleration towards the Earth would, by the law of gravitation, be about 3600 times its present acceleration.

Now A being the mean radius of the Moon's orbit, and T the periodic time (about $27\frac{1}{3}$ days), the acceleration of the Moon towards the Earth is $\frac{4\pi^2 A}{T^2}$, supposing that the mass of the Moon is small compared with that of the Earth.

Substituting the known values of A and T, this acceleration is ·2758 centimetre-second units.

If this is multiplied by 3600, we obtain 978·6 cm. sec. units, which is almost exactly the acceleration of a falling body at the Earth's surface.

Hence we ascribe the acceleration of falling bodies at the Earth's surface to terrestrial gravitation.

§ 18. Variations of g at the Earth's Surface.

If the Earth were a sphere at rest, the acceleration g of a freely falling body would be the same in magnitude everywhere on the Earth's surface. The value of g, however, is different in different places, being affected by the following circumstances :—(1) the deviation of the Earth's surface from a spherical form, (2) the height of the place of experiment above sea-level, (3) the Earth's rotation on its polar axis.

(1) The figure of the Earth is not accurately spherical; it resembles the surface formed by the rotation of an ellipse about its minor axis, for the equator is a circle of diameter 7927 miles, and the polar diameter is 7899 miles.

Hence the distance of a body from the Earth's centre increases in passing along the surface from the pole to the equator, and g is greatest at the pole and least at the equator.

(2) For the same reason g diminishes as the height above sea-level increases.

(3) We have considered g as an acceleration relatively to the Earth's centre; it is, however, really an acceleration relatively to the Earth's surface in our locality.

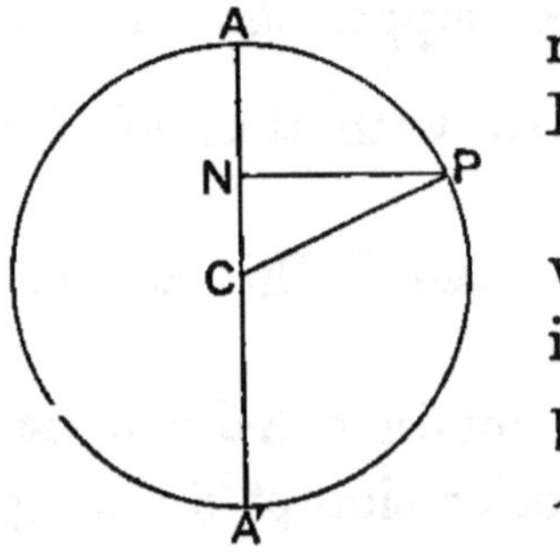

Fig. 105.

Now the Earth rotates with angular velocity ω on its axis AA'. If PN is the perpendicular distance of a point P on the Earth's surface from AA', $\omega^2 PN$ is the acceleration of P relatively to the Earth's axis.

Reversing this, we find that the acceleration of a falling body relatively to P is compounded of a, the true acceleration along CP relative to the Earth's centre, and $\omega^2 PN$ directed along NP.

In estimating $\omega^2 PN$ we may regard the Earth as a sphere of radius R.

If λ is the latitude of P, $PN = R \cos \lambda$, and the component of $\omega^2 PN$ along PC is $\omega^2 R \cos^2 \lambda$.

Therefore the true acceleration along PC is

$$a - \omega^2 R \cos^2 \lambda.$$

The effect of the Earth's rotation is therefore to diminish the value of g by an amount which is greatest at the equator and least at the poles.

Taking $R = 4000$ miles $= 21{,}120{,}000$ feet, and

$$\omega = \frac{2\pi}{86164},$$

we find that at the equator g is diminished by about $\frac{1}{289}$th of its value, in consequence of the Earth's rotation.

Value of g within the Earth.

If the density of the Earth were uniform, the value of g within the Earth would be proportional to the distance from the Earth's centre. At the bottom of a coal shaft, $\frac{1}{2}$ mile deep, the value of g would be $\frac{1}{8000}$th part less than it is at the surface.

Experiments with the pendulum, conducted by Sir George Airy, have however shown that g is greater at the bottom of the shaft than at the surface.

We can account for this by supposing that the density of the Earth increases with the depth below the surface, and this hypothesis is justified, since the mean density of the Earth—determined by a method to be described later—much exceeds the mean density of the rocks, &c., at the surface.

Let M be the mass of the Earth, ρ its average density, g the acceleration of gravity at the surface, g' the acceleration at a depth h, ρ' the density of the nucleus of radius $R-h$.

Then $$g=\frac{M}{R^2}=\frac{4}{3}\pi\rho R.$$

And $$g'=\frac{4\pi}{3}\rho'(R-h).$$

Therefore $g'>g$ if $\rho'(R-h)>\rho R$.

And $\frac{h}{R}$ being less than $\frac{1}{8000}$, $g'>g$ if $\frac{\rho'}{\rho}>\frac{8000}{7999}$.

§ 19. It follows from the law of gravitation that bodies on the Earth attract each other, and experiments arranged on a suitable scale show this.

The following method was devised by Jolly.

P, Q are the pans of a delicate balance.

A glass globe filled with mercury is placed in Q, and balanced by a mass m in the pan P.

A sphere of lead, 1 metre in diameter, is now placed vertically below the middle of the pan Q. Under its attraction the pan Q descends and the beam can only be brought to its former position of rest by the addition of a mass ∂ to m.

The globe of mercury in the pan Q is then so much the larger portion of the mass there that ∂g may be taken as the force exerted by the sphere on the globe of mercury.

By this experiment the mass of the Earth may be determined in grams.

Let M, m, μ be the masses (measured in grams) of the Earth, the lead sphere, and the globe of mercury, R the radius of the Earth, r the distance between the centre of the globe and sphere.

Then since $g = \frac{kM}{R^2}$, and $\partial g = \frac{km\mu}{r^2}$,

$$M = \frac{m\mu}{\partial} \frac{R^2}{r^2}.$$

In the experiment
$m = 57\cdot7$ kilos.
$\mu = 5$ kilos.
$\partial = \cdot589$ mg.
$R = 6366$ kiloms.
$r = \cdot5686$ m.

M can be determined from these data, and thence the Earth's density, which is $\frac{M}{\frac{4}{3}\pi R^3}$, or approximately 5·7.

Hence (§ 10) the masses of the sun and the planets can be found, and thence the densities of these bodies can be deduced, since their radii are known.

CHAPTER VII.

ELASTICITY.

§ 1. Stress and Strain.

Let a body be maintained in equilibrium by forces P, Q, R applied at points A, B, C of its surface.

Since a portion of the body, $EGFAE$, containing the point A, but not B and C, is also in equilibrium, other forces must act on this portion, and their resultant must be equal and opposite to P. These forces can only be supplied by the action of the other part of the body, $EBCFE$; since if this part were removed $EGFAE$ would not remain in equilibrium.

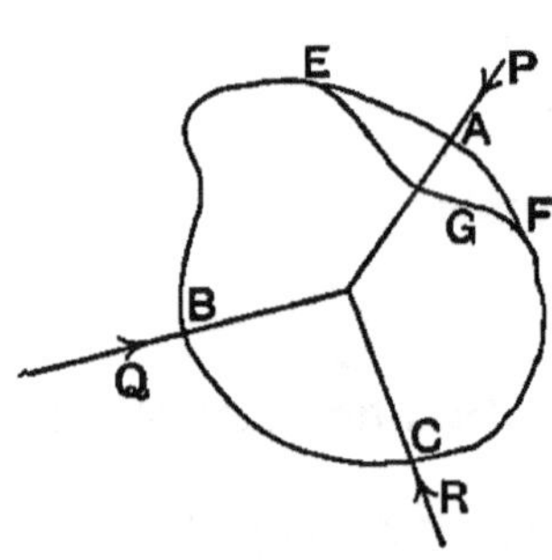

Fig. 106.

Similarly, if we take any other surface within the body, the two portions of matter separated by the surface will generally exert force on one another. The forces at different points of the surface are not necessarily the same either in magnitude or in direction. These forces, regarded as mutual actions of the parts of a body, are called Stresses.

If S be the area of a small closed curve on a plane X within the body, P a point within the curve S, and F the total force which the matter on one side of S exerts across S

on the remaining matter, then F is the Total Stress across S, and $\frac{F}{S}$ is the Stress at P across the plane on which S lies.

The Stress is called a Tension or Pressure according as the force F is a pull or thrust, and it is uniform if it has the same value at different points of the plane.

If the Stress across the plane X is not uniform, the area S enclosing the point P must be made very small; the limit to which $\frac{F}{S}$ approaches when S is indefinitely small is then the Stress at P.

Examples.

1. A wire of 1 mm. radius is stretched by the weight of 2 kilograms. To find the stress across any cross-section, neglecting the weight of the wire.

Here
$$S = \pi \times (\cdot 1)^2$$
$$P = 2000 \times 981.$$

This is the Total Stress across a section.

Therefore the stress is $\frac{P}{S}$ or $\frac{2 \times 981 \times 10^6}{\pi}$.

2. A cylinder of radius r, and density ρ, is suspended with its axis vertical, its upper end being fixed. Find the stress across a horizontal section of the rod, distant d from the lower end.

Here P is the weight of the cylinder of length d.

$$\therefore \quad P = \pi d r^2 \rho g,$$
$$\text{and} \quad S = \pi r^2.$$

The stress is $\frac{P}{S}$ or $\rho d g$.

If the external forces which maintain a body in equilibrium are doubled, the equilibrium is not disturbed; it thus appears that the stresses in the body are doubled too.

It is found that alterations in the internal stresses are accompanied by small changes in the form and size of the body, which are called Strains.

The branch of Mechanics which treats of the relation between stresses and strains is called Elasticity. Before investigating this relation, we must find some way of classifying and measuring strains.

§ 2. Plane Strain.

Let us consider the simplest changes in the form of the parts of a plane figure which is strained so that it remains plane.

Let M and m be two precisely similar plane membranes, stretched by similar forces applied at their boundaries, S any curve drawn on the membrane M, s the curve on m which coincides with S, when m is superposed on M so that the membranes coincide.

If the tensions at the boundary of m are altered, m will assume another form μ with a new boundary, and s will assume a form which we may call σ. We have then a figure M and a curve S precisely similar to the forms which μ and σ had before strain; we may therefore regard μ and σ as the strained forms of M and S, a point in μ corresponding to each point in M.

It is clear that we may take any plane curve for S, and make σ assume an indefinite variety of forms. We shall, however, only take the case when S is a straight line, and the corresponding curve σ is a straight line too. Then if A and B are points in M, and α and β the corresponding points in μ, a point γ in $\alpha\beta$ corresponds to a point G in AB.

Let us further assume that $AG : GB :: \alpha\gamma : \gamma\beta$.

Since AB is changed into $\alpha\beta$ by strain, $\dfrac{\alpha\beta - AB}{AB}$ is called the elongation of AB, and since this fraction is equal to $\dfrac{\alpha\gamma - AG}{AG}$, the elongation of AG is the same as that of AB.

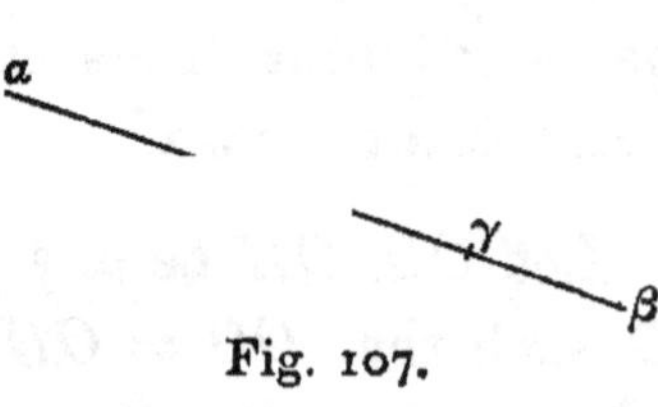

Fig. 107.

The elongation of a line is its proportional increase of length, i.e. its increase of length divided by its original length, and the assumption that $\dfrac{AG}{GB} = \dfrac{\alpha\gamma}{\gamma\beta}$ is equivalent to making the elongation of all parts of AB the same. This strain, being the same at all points, is said to be homogeneous.

If $AG = GB$, then $\alpha\gamma = \gamma\beta$.

§ 3. Equal and parallel straight lines are equal and parallel after strain.

Let AB, CD be equal and parallel straight lines; $\alpha\beta$, $\gamma\delta$ their positions after strain. Then AD, BC bisect one another, and

$$AE = ED, \ BE = EC.$$

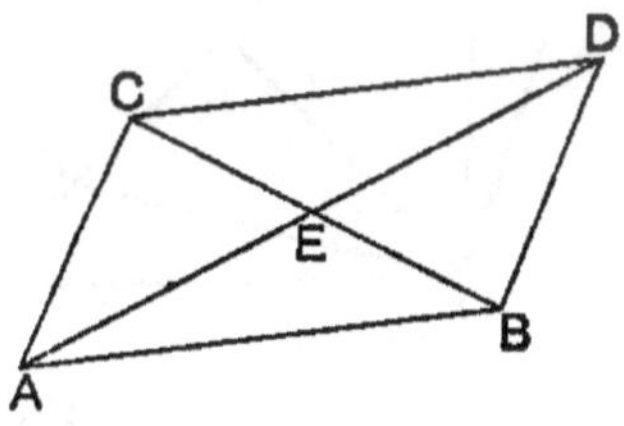

But $\alpha\delta$, $\beta\gamma$ are the strained positions of AD, BC, and ϵ is the strained position of E. Therefore by the last section

$$\beta\epsilon = \epsilon\gamma, \ \alpha\epsilon = \epsilon\delta.$$

Hence (Euclid I. 4)

$$\gamma\delta = \alpha\beta,$$

and the angles $\gamma\delta\alpha$, $\delta\alpha\beta$ are equal.

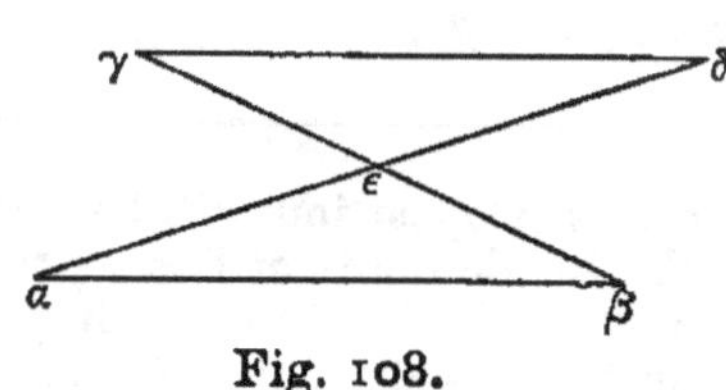

Fig. 108.

Therefore $\alpha\beta$, $\gamma\delta$ are equal and parallel.

A parallelogram strains into a parallelogram, but its angles are generally altered by strain. Thus a right angle generally becomes an oblique angle.

§ 4. To prove that through any point *O* a pair of perpendicular lines can be drawn which are perpendicular after strain [1].

Let OG, OH be any pair of perpendicular lines through O, such that $OG = OH$; og, oh the position of these lines after strain; E, e the middle points of GH, gh.

Then e is the strained position of E.

Let OP, OQ be another pair of lines through O; op, oq

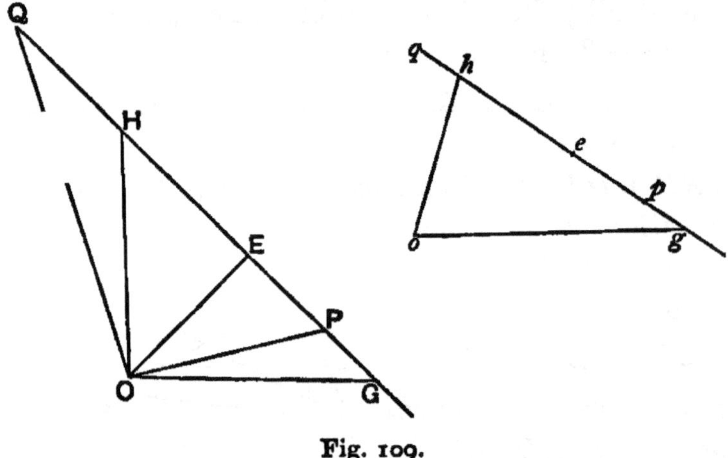

Fig. 109.

their strained positions; then QOP is a right angle if $ep \,.\, eq = eg^2$.

---

[1] The proposition will be obvious to the reader who is acquainted with the elements of Descriptive Geometry. For a strained figure is evidently similar to an orthogonal projection of the unstrained figure. Hence to a pencil in involution in the unstrained figure corresponds an involution in the strained figure, and to a right-angled involution corresponds an overlapping involution which has one pair of rays rectangular, and only one, unless all are rectangular.

$$\text{For} \quad \frac{ep}{EP} = \frac{eq}{EQ} = \frac{eg}{EG}.$$

$$\text{Therefore} \quad \frac{ep \cdot eq}{EP \cdot EQ} = \frac{eg^2}{EG^2}.$$

And if $ep \cdot eq = eg^2$; $EP \cdot EQ = EG^2 = EO^2$.

Therefore $EP : EO :: EO : EQ$.

And the angles PEO, OEQ are equal.

Therefore PEO, OEQ are similar triangles, and since the angles EPO and EOP are respectively equal to EOQ and EQO, the angle POQ is equal to the sum of OPQ and OQP, and is a right angle.

We have now to find two points a and b on gh which subtend a right angle at o, and are such that $ea \cdot eb = eg^2$.

Describe the circle ohg meeting oe produced in f.

Bisect of in k, and draw kl perpendicular to of meeting hg in l. With centre l and radius lo (or lf) describe the circle oaf meeting gh in a and b.

Then

$$ea \cdot eb = eo \cdot ef = eg^2,$$

and aob is a right angle, for it is in a semi-circle.

Therefore a and b are the points required.

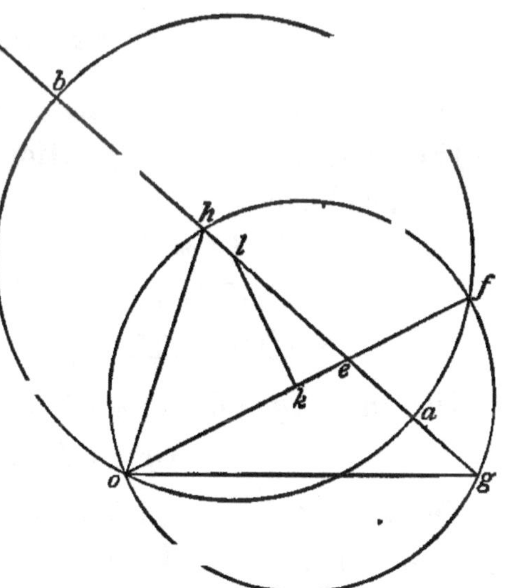

Fig. 110.

The straight lines OA, OB, of which oa, ob are the strained positions, are the *axes* of the strain, and their elongations are the *principal elongations*.

If an axis is shortened by strain its elongation is negative.

§ 5. Plane Strain referred to its axes.

Let OA, OB be the principal axes of the strain, oa, ob the lines into which OA, OB are strained; then if the principal elongations ϵ_1, ϵ_2 are known, the strained position of any other point P can be determined.

For draw PM, PN perpendicular to OA, OB respectively. Make $Om = OM(1+\epsilon_1)$, and $On = ON(1+\epsilon_2)$.

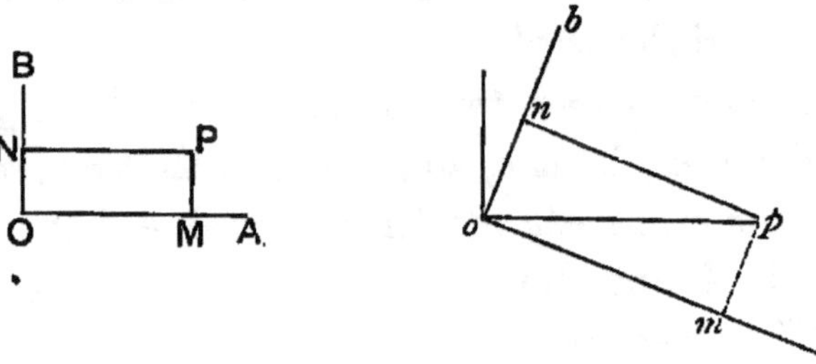

Fig. 111.

Complete the rectangle $nomp$, and let $\epsilon_2 < \epsilon_1$.

Then p is the strained position of P.

We also have

$$\begin{aligned} Op^2 &= Om^2 + mp^2 \\ &= OM^2(1+\epsilon_1)^2 + MP^2(1+\epsilon_2)^2, \\ &> (OM^2 + MP^2)(1+\epsilon_2)^2 \text{ since } \epsilon_2 < \epsilon_1. \end{aligned}$$

Therefore $Op > OP(1+\epsilon_2)$, and similarly

$$Op < OP(1+\epsilon_1).$$

Therefore the elongation of any line in the plane is greater than ϵ_2, and less than ϵ_1. Or, one principal elongation is the greatest and the other the least of all elongations.

When oa, ob are parallel to OA, OB respectively, the strain is said to be pure.

By a rotation of *oab* about *o* without change of form, *oa* can be made parallel to *OA*, and *ob* to *OB*.

Hence any plane strain reduces either to a pure strain, or to a pure strain and a rotation.

Uniform Extension.

If $\epsilon_1 = \epsilon_2$, $op = OP\ (1 + \epsilon_1)$, and $\dfrac{om}{mp} = \dfrac{OM}{MP}$.

Therefore the angles *POM*, *pom* are equal.

In this case, corresponding angles in the strained and unstrained figures are equal, and corresponding lines are in a constant ratio; the figures are similar to one another in all their parts, only differing in the scale on which they are drawn.

This strain may be called a uniform areal extension.

The Strain-Ellipse.

Let the strained axes *oa*, *ob* be superposed on the unstrained axes, *OA*, *OB*.

Take any point *P* in the unstrained figure, and with centre *O* and radius *OP* describe a circle. Make

$Op' = OP\,(1 + \epsilon_1)$

and describe a circle with centre *O* and radius Op'; draw $p'n$ perpendicular to *OA* and take *p* in $p'n$ such that

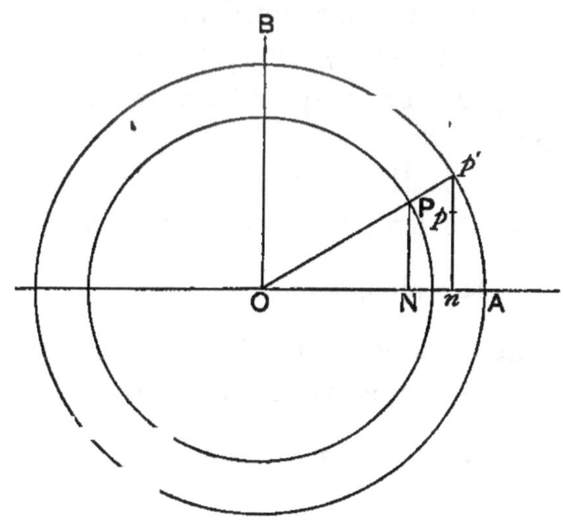

Fig. 112.

$\dfrac{pn}{p'n} = \dfrac{1 + \epsilon_2}{1 + \epsilon_1}$. Draw *PN* perpendicular to *OA*.

Then if $OP = a$, p lies on an ellipse, the semi-axes of which are $a(1+\epsilon_1)$, $a(1+\epsilon_2)$.

And $\frac{On}{ON} = \frac{p'n}{PN} = \frac{Op'}{OP} = 1 + \epsilon_1$.

And $\frac{pn}{PN} = \frac{pn}{p'n} \cdot \frac{p'n}{PN} = 1 + \epsilon_2$.

Therefore p is the strained position of P.

Therefore a circle strains into an ellipse with axes along Oa and Ob, the strained positions of the axes of strain.

§ 6. The propositions on Plane Strain that remain to be discussed are only true when the squares of the elongations are neglected. This approximation is not only convenient but necessary in the present state of our knowledge, as the following considerations may show.

The law connecting stresses and the corresponding strains was laid down by Hooke, and is as follows:—

Strain is proportional to the corresponding stress.

Let us apply this to the case of a cylindrical wire hanging vertically and stretched by a weight at its lower end.

If l is the length of the wire when unstretched, and $l+d$ its length when a weight P is suspended from it, $\frac{d}{l}$ is the elongation.

Let a second weight P be suspended from the wire; if the elastic quality is not altered by strain the wire will be lengthened by d' where $\frac{d'}{l+d} = \frac{d}{l}$,

or
$$d' = d + \frac{d^2}{l}.$$

The total elongation due to the weight $2P$ is $\frac{2d}{l} + \frac{d^2}{l^2}$.

But, according to the law, if a weight P produces an elongation $\frac{d}{l}$, a weight $2P$ produces an elongation $2\frac{d}{l}$.

Hence if we assume Hooke's law, we must regard $\frac{d^2}{l^2}$ as negligible; now experiment proves that Hooke's law is in good accordance with the simpler properties of elastic bodies, and that $\frac{d^2}{l^2}$ is practically insensible.

We may therefore neglect the squares of elongations and the products of different elongations.

§ 7. Change of area by Strain.

In figure 111, the area

$$ompn = Om \cdot On = OM \cdot ON(1+\epsilon_1)(1+\epsilon_2).$$

The proportional change of area or areal extension is

$$\frac{ompn - OMPN}{OMPN} \text{ or } (1+\epsilon_1)(1+\epsilon_2)-1,$$

or $\epsilon_1+\epsilon_2$ since $\epsilon_1\epsilon_2$ is negligible.

Any other strained area can be divided into rectangles with their sides parallel to oa and ob.

Therefore $\epsilon_1+\epsilon_2$ is the areal extension of any figure on the plane.

Shearing Strain.

Let $\epsilon_1+\epsilon_2=0$, or $\epsilon_1=-\epsilon_2=\epsilon$.

Then if $PQRS$ be a square with its sides parallel to the axes of the strain, it becomes by strain a rectangle $pqrs$ such that $pq=PQ(1+\epsilon)$, and $qr=QR(1-\epsilon)$.

Then

$$pr^2 = pq^2+qr^2 = PQ^2(1+\epsilon)^2+QR^2(1-\epsilon)^2,$$

$$= PQ^2\{(1+\epsilon)^2+(1-\epsilon)^2\} = 2PQ^2, \text{ since } \epsilon^2 \text{ is neglected,}$$

$$= PR^2.$$

Therefore the diagonals of the square are not altered in length by the strain.

Let the strained and unstrained figures be superposed, so that PQ, pq are parallel, and the point of intersection of PR, QS coincides with that of pr, qs at O. Then the same circle with centre O circumscribes $pqrs$ and $PQRS$.

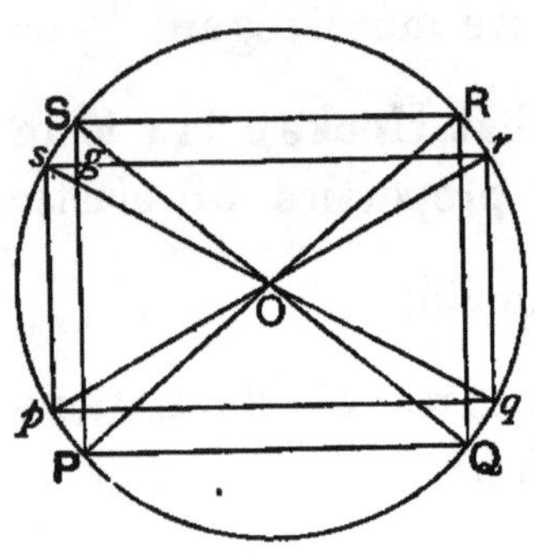

Fig. 113.

Let PS, rs intersect in g; then if $PS = 2a$, $Sg = a\epsilon = sg$.

And $sS^2 = sg^2 + gS^2 = 2a^2\epsilon^2$,

or $sS = a\epsilon\sqrt{2}$.

And since $Os = a\sqrt{2}$, $\dfrac{sS}{Os} = \epsilon$.

Now $sOS + pOP$ is the change in the angle between the diagonals of the square, and $sOS = pOP$.

Therefore $\epsilon, -\epsilon$ being the principal elongations, 2ϵ is the change in the angle between the diagonals of the square.

This strain is called a shearing strain; it consists in a distortion of the figure without appreciable change of size. We may also regard this strain from another point of view.

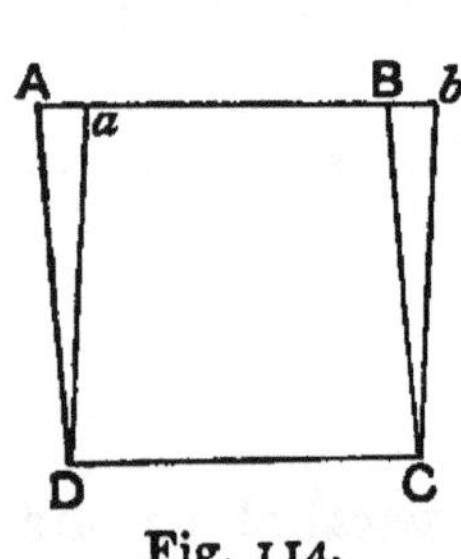

Fig. 114.

If $ABCD$ is a square the sides of which are parallel to PR, QS, the sides and area of $ABCD$ are not altered by the strain, but the angles of the square become oblique.

The square after strain becomes the rhombus $aDCb$; aD differing from AD by a quantity of order ϵ^2, which we neglect.

§ 8. Superposition of Strains.

We shall denote a strain in which the principal elongations are ϵ_1, ϵ_2 by $[\epsilon_1, \epsilon_2]$.

Let ϵ_1, ϵ_2 be the principal elongations in a strained figure, and let the figure receive a further strain $[\epsilon_1', \epsilon_2']$ with the same axes as $[\epsilon_1, \epsilon_2]$.

Let P be a point in the unstrained figure, and let oa lie on OA, and ob on OB.

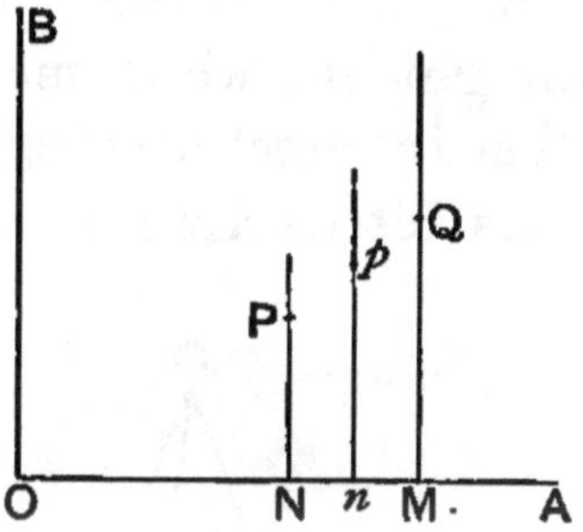

Draw PN perpendicular to OA and make $On = ON(1+\epsilon_1)$.

From n draw np perpendicular to OA, and make $np = NP(1+\epsilon_2)$.

Then p is the position of P after the strain $[\epsilon_1, \epsilon_2]$.

Fig. 115.

Make $OM = On(1+\epsilon_1')$.

Draw MQ perpendicular to OA, and make

$$MQ = np(1+\epsilon_2').$$

Then Q is the position of p after the strain $[\epsilon_1', \epsilon_2']$, and the position of P after the strains $[\epsilon_1, \epsilon_2]$ and $[\epsilon_1', \epsilon_2']$.

$$OM = On(1+\epsilon_1') = ON(1+\epsilon_1)(1+\epsilon_1')$$

$$= ON(1+\epsilon_1+\epsilon_1'), \text{ since } \epsilon_1\epsilon_1' \text{ is negligible.}$$

Similarly $\qquad MQ = NP(1+\epsilon_2+\epsilon_2').$

Thus the strained position of P is the same as that due to a single strain $\epsilon_1+\epsilon_1'$, $\epsilon_2+\epsilon_2'$.

A strain $[\epsilon_1, \epsilon_2]$ is equivalent to the two strains

$$\left[\frac{\epsilon_1+\epsilon_2}{2}, \frac{\epsilon_1+\epsilon_2}{2}\right] \text{ and } \left[\frac{\epsilon_1-\epsilon_2}{2}, -\frac{\epsilon_1-\epsilon_2}{2}\right],$$

i. e. to an areal dilatation superposed on a shearing strain.

§ 9. **Strain of a solid body.**

Let us assume, as before, that straight lines strain into straight lines and that the elongations of all parts of the same straight line are the same. This strain is called homogeneous.

A plane is the surface traced out by an indefinite straight line which moves so that it always intersects two other intersecting straight lines.

As intersecting straight lines strain into intersecting

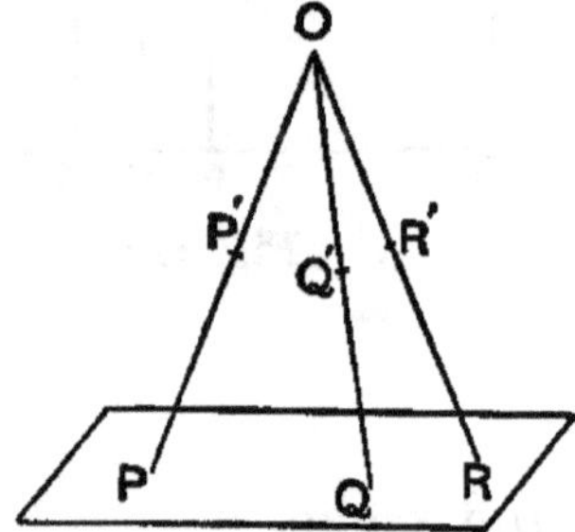

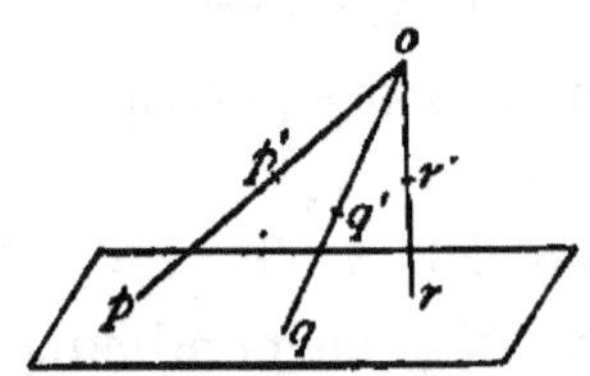

Fig. 116.

straight lines, a plane strains into another plane.

Let PQR be a plane in the unstrained figure, OP, OQ, OR three lines which are not in the same plane, and let $\frac{OP'}{OP} = \frac{OQ'}{OQ} = \frac{OR'}{OR}$.

Then (Euclid XI. 17), P', Q', R' lie in a plane parallel to PQR. Now let pqr be the plane into which PQR strains, and let op, oq, or be the strained positions of OP, OQ, OR, p', q', r' the strained positions of P', Q', R'.

Then $$\frac{op'}{op} = \frac{OP'}{OP} \text{ and } \frac{oq'}{oq} = \frac{OQ'}{OQ}.$$

Therefore $$\frac{op'}{op} = \frac{oq'}{oq} = \frac{or'}{or}.$$

And p', q', r' lie in a plane parallel to pqr.

Hence parallel planes strain into parallel planes.

Let O be the centre of a sphere in the unstrained figure. The section of the sphere by any plane is a circle, and it has been proved that by strain the circle becomes an ellipse.

All radii of the sphere are elongated or contracted to a greater or less extent by strains, and there is at least one radius OA whose elongation is greater, and another OC whose elongation is less than that of any other radius. If oa, oc are the strained positions of these radii, oa is greater and oc less than any other strained radius in the plane oac. Therefore oa and oc are perpendicular to one another, and the elongations of OA, OC are principal elongations.

Draw ob perpendicular to the plane aoc.

Then oc is less than any other elongation in the plane boc, and consequently is a principal elongation in this plane; therefore ob is the other principal elongation. Similarly ob is a principal elongation in the plane aob. Therefore if OB be the unstrained position of Ob, the angles AOB, BOC, COA are right angles, and strain into right angles.

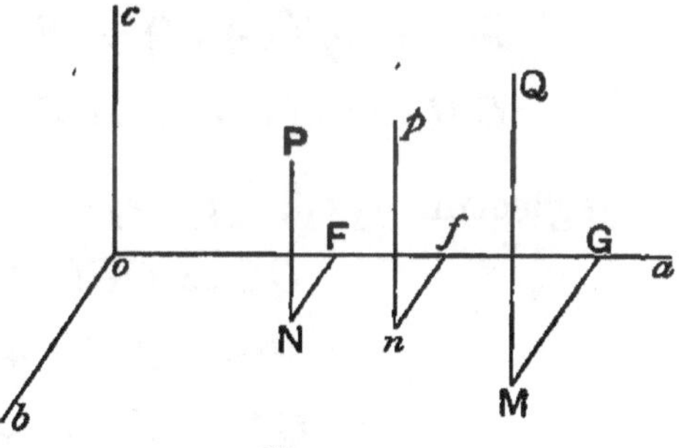

Fig. 117.

And a cube, the sides of which are parallel to OA, OB, OC, strains into a rectangular parallelopiped.

OA, OB, OC are called the axes of the strain, and the elongations ϵ_1, ϵ_2, ϵ_3 along them are called the principal elongations.

Let the strained and unstrained figures be superposed so that oa, ob, oc lie on OA, OB, OC.

From any point P draw PN perpendicular to the plane aob meeting it in N, and from N draw NF perpendicular to oa.

Make $Of = OF.(1+\epsilon_1)$. Draw fn parallel to FN, and make $fn = FN(1+\epsilon_2)$.

Draw np parallel to NP, and make $np = NP(1+\epsilon_3)$.

Then p is the strained position of P.

The cubical dilatation is $(1+\epsilon_1)(1+\epsilon_2)(1+\epsilon_3)-1$ or $\epsilon_1+\epsilon_2+\epsilon_3$, when products of elongations are neglected.

§ 10. Superposition of Strains.

Let the strained figure be further strained with elongations $(\epsilon_1', \epsilon_2', \epsilon_3')$. Then if Q be the strained position of p, and QM, MG be parallel to pn, nf respectively,

$$OG = Of(1+\epsilon_1') = OF(1+\epsilon_1')(1+\epsilon_1),$$

$$MG = nf(1+\epsilon_2') = NF(1+\epsilon_2')(1+\epsilon_2),$$

$$QM = pn(1+\epsilon_3') = PN(1+\epsilon_3')(1+\epsilon_3).$$

Neglecting $\epsilon_1\epsilon_1'$, $\epsilon_2\epsilon_2'$, $\epsilon_3\epsilon_3'$,

$$OG = OF(1+\epsilon_1+\epsilon_1'),$$

$$MG = NF(1+\epsilon_2+\epsilon_2'),$$

$$QM = PN(1+\epsilon_3+\epsilon_3').$$

And Q is the position which P would assume after a strain whose principal elongations are

$$\epsilon_1+\epsilon_1', \ \epsilon_2+\epsilon_2', \ \epsilon_3+\epsilon_3'.$$

Shearing Strain.

Let $\epsilon_3 = 0$ and $\epsilon_1 = -\epsilon_2 = \epsilon$.

Then if OX, OY are the external and internal bisectors of the angle OAB in the unstrained figure, the lines OX, OY

are unaltered by strain, but the angle YOX is altered by 2ϵ, CO remaining perpendicular to OX and OY.

Since YOX is the angle between the planes COX, COY, the change in this angle is called the shear of the planes.

Since $\epsilon_1 + \epsilon_2 + \epsilon_3 = 0$, there is no cubical dilatation in this strain.

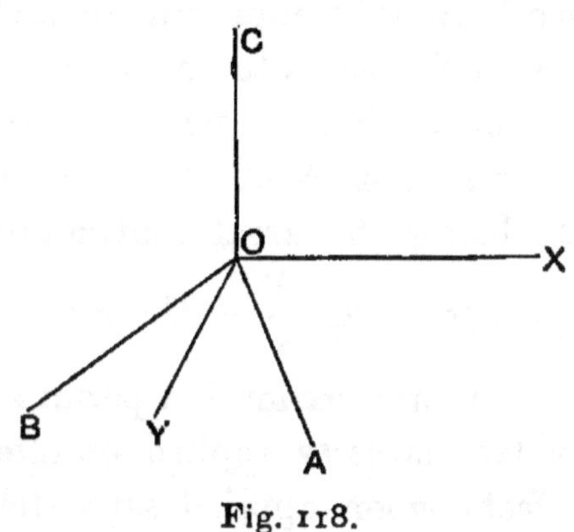

Fig. 118.

Uniform Dilatation.

If $\epsilon_1 = \epsilon_2 = \epsilon_3 = \epsilon$, all elongations are equal; this strain is a uniform cubical dilatation of magnitude 3ϵ.

Reduction of Strain.

Any strain can be reduced to uniform cubical dilatations and shears.

Denote a strain whose principal elongations are $\epsilon_1, \epsilon_2, \epsilon_3$ by $[\epsilon_1, \epsilon_2, \epsilon_3]$.

Then a simple elongation ϵ_1 is $[\epsilon_1, 0, 0]$, and by the principle of superposition of strains it is equal to

$$\left[\frac{\epsilon_1}{3}, \frac{\epsilon_1}{3}. \frac{\epsilon_1}{3}\right] + \left[\frac{\epsilon_1}{3}, -\frac{\epsilon_1}{3}, 0\right] + \left[\frac{\epsilon_1}{3}, 0, -\frac{\epsilon_1}{3}\right].$$

But the first term on the right-hand represents a cubical dilatation and the two others represent shears.

Hence a simple elongation is equivalent to a uniform cubical dilatation and two shears, and the most general strain considered is made up of three elongations at right angles to one another, and can therefore be reduced to cubical dilatations and shears.

§ 11. By the action of stress the position and area of a surface across which stress is exerted are changed, and in estimating stress as a force per unit area strict accuracy would be obtained by referring to the strained state of the body. It has, however, been already mentioned that our work is only approximate, and it thence appears that we may refer stress to the unstrained state of the body.

Thus, when a wire is stretched by a weight P, the cross-section S of the wire is diminished to $S(1-2\epsilon)$, 2ϵ being the areal contraction of the cross-section; we, however, take $\frac{P}{S}$ as the stress, and not $\frac{P}{S(1-2\epsilon)}$.

This approximation permits us to assume that the effect of two stresses applied simultaneously is the sum of their effects when applied separately, so long as the resulting strain is small.

If the strain of the solid is everywhere the same, the stress is the same at all points.

Shearing Stress.

If equal pressures T are applied perpendicularly to the faces *be*, *og* of the cube *oadbfcga*, *ob* and the edges parallel to it will contract, and all lines perpendicular to *ob* will be equally elongated.

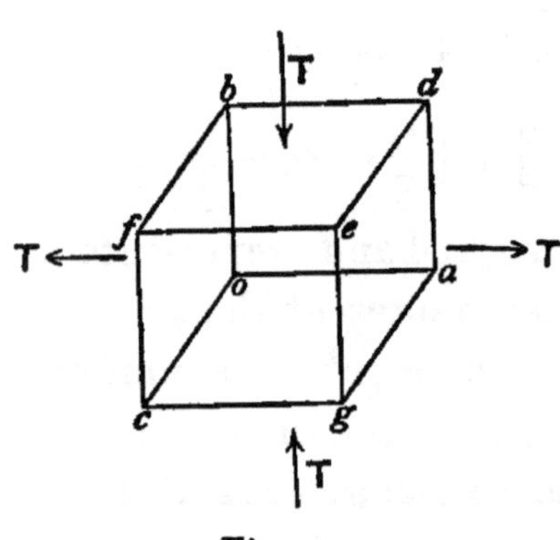

Fig. 119.

Let a be the elongation of *oa* and *oc*, $-\beta$ the elongation of *ob*.

If equal tensions T are applied perpendicularly to the faces *of* and *ae*, $-a$ is the elongation of *ob* and *oc*, and β is the elongation of *oa*, due to the tensions T.

Superposing the two stresses, we find that *oc* has no elongation, *oa* and *ob* have elongations $\alpha+\beta$, $-(\alpha+\beta)$.

Hence the stress produces shear only, and is called a Shearing Stress.

Hydrostatic Stress.

If equal tensions T are applied perpendicularly to the six faces of the cube, the resulting strain is the sum of the strains due to each pair of tensions; therefore each edge of the cube is elongated by $\beta-2\alpha$, and the angles of the figure are unaltered by strain. Hence the elongations are principal elongations, and the strain is a cubical dilatation without change of form. The corresponding stress may be called a Hydrostatic Stress.

Transformation of Shearing Stress.

Let the cube *oadbfcga* be cut in two by the plane *agfb*, forming a prism on a triangular base *fcg*; then if no external forces (as gravity) act on the substance of the prism, its equilibrium is due to the stresses on its faces.

Let the triangular faces be free from stress, and let normal stresses T, $-T$ be applied to the faces *ca*, *cb*. Since the stress on each face is uniformly distributed, it can be replaced by the total stress acting at the centroid of the face.

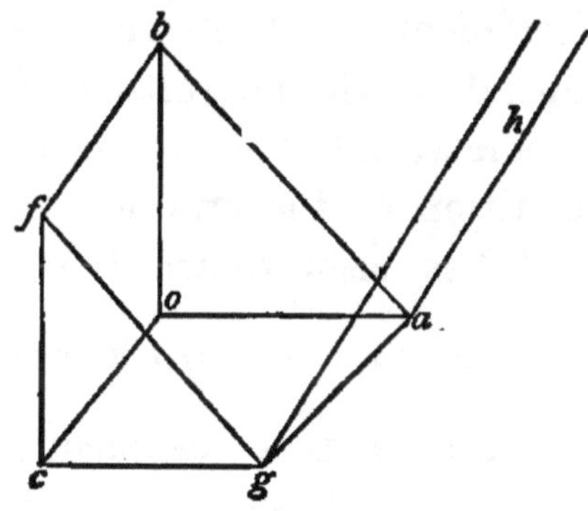

Fig. 120.

If $oa = l$, the prism is in equilibrium under forces l^2T parallel to *ao*, l^2T parallel to *ob*, and the total stress on the plane *abfg*. This latter force must therefore be

$Tl^2\sqrt{2}$, parallel to ba, and it is distributed over the area $l^2\sqrt{2}$.

Therefore the stress on the plane $abfg$ is T, and it is parallel to ba.

If a plane gah is drawn through ag perpendicular to afg, the stress on this plane is tangential to it, and is of magnitude T.

Relation between Stress and Strain. Hooke's Law.

The relation between Stresses and the corresponding small Strains was first enunciated by Hooke in 1676, and may be stated as follows:—

Strains are proportional to the Stresses which produce them.

The principal stresses that we have to consider are hydrostatic stress and shearing stress.

Let P be a hydrostatic stress, producing a uniform dilatation δ. By Hooke's law $\frac{P}{\delta}$ is the same for all values of P, and if $P = k\delta$, k is called the Elasticity of bulk, or the Coefficient of Resistance to Compression. It does not depend on the magnitude of the pressure so long as this is moderate, but it is determined by the nature and physical condition of the strained body.

It has been shown that a shearing stress S produces a shear 2ϵ. By Hooke's law $\frac{S}{2\epsilon}$ is a constant, and if $S = 2n\epsilon$, n is a coefficient, which depends only on the nature and physical condition of the body which undergoes shear.

n is called the Coefficient of Rigidity of the body, or the Elasticity of Form.

§ 12. Reduction of any stress to hydrostatic and shearing stress.—Young's Modulus.

In general the stress within a body may alter both the form and size of the parts of the body. But any strain can be reduced to cubical dilatation δ and shearing strain s, and therefore the stress consists of a hydrostatic stress $k\delta$ superposed on a shearing stress $2ns$. Hence any stress reduces to a combination of shearing and hydrostatic stresses.

An example will make this point clearer.

Let a wire hanging vertically with its upper end fixed be stretched by a weight W attached to its lower end. Neglect the weight of the wire, and let l, S be its length and the area of its cross-section before stretching, $l + \lambda$ the length after stretching.

Then since $\frac{\lambda}{l}$ is the elongation ϵ, and $\frac{W}{S}$ is the stress,

$\frac{W}{S} = \mu \frac{\lambda}{l}$, where μ does not depend on W, l, or S.

The coefficient μ is the measure of the resistance which the wire offers to stretching, and is called Young's Modulus.

Take a small cube in the wire with its faces of, $o'c$ horizontal.

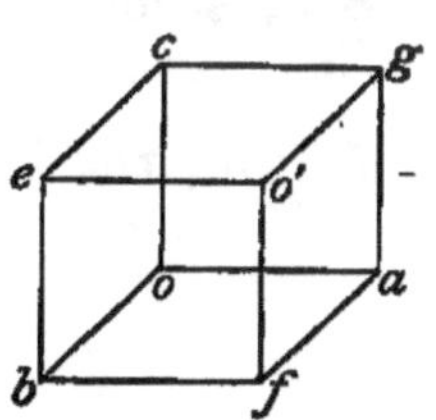

Fig. 121.

The stress across these faces is $\frac{W}{S}$, which we denote by T, and there is no stress across the other faces.

This stress is equivalent to tensions $\frac{T}{3}$ across all six faces, tensions $\frac{T}{3}$ and $-\frac{T}{3}$ across of, oe respectively, and the opposite faces, and tensions $\frac{T}{3}$ and $-\frac{T}{3}$ across *of*, *og* and the opposite faces.

The equal tensions across all faces of the cube form a hydrostatic stress $\frac{T}{3}$ and the corresponding strain is $\frac{T}{3k}$, or a linear dilatation $\frac{T}{9k}$.

The tensions $\frac{T}{3}$, $-\frac{T}{3}$ across *of*, *oe* and the opposite faces form a shearing stress. The corresponding principal elongations are along *oc* and *oa* and are $\frac{T}{6n}$ and $-\frac{T}{6n}$.

The tensions across *of*, *og* and the opposite faces give similar elongations.

Hence on the whole the elongation of the wire is given by

$$\epsilon = T\left\{\frac{1}{9k} + \frac{1}{3n}\right\}.$$

And since $$T = \mu\epsilon, \quad \mu = \frac{9kn}{3k+n}.$$

Again, the lateral linear contraction of the wire is

$$T\left\{\frac{1}{6n} - \frac{1}{9k}\right\},$$

and as the wire cannot expand laterally by stretching, k cannot be less than $\frac{2}{3}n$.

Modulus of elasticity for simple longitudinal strain λ.

This strain is equivalent to a cubical dilatation λ and two shears.

The hydrostatic stress is a tension $k\lambda$ on all planes, and if *oc* be the direction of the elongation, the stress corresponding to the first shear is a tension $\frac{2n\lambda}{3}$ across of, and parallel planes, and $-\frac{2n\lambda}{3}$ across oe, and parallel planes. The stress corresponding to the second shear is similarly formed.

Therefore $\left(k+\frac{4n}{3}\right)\lambda$ is the longitudinal stress (across of, and it is accompanied by equal lateral stresses $\left(k-\frac{2n}{3}\right)\lambda$ across all planes parallel to the direction of elongation.

§ 13. Torsion.

A straight uniform wire is suspended vertically from a fixed support, and carries a horizontal bar attached to its lower end.

The bar is deflected from its position of rest by equal and opposite forces P applied to its ends perpendicular to its length, and comes to rest when it makes a certain angle ψ with its former position.

The lower end of the wire is thus twisted through an angle ψ, and the wire is said to have a twist, or to undergo Torsion.

If S is any horizontal section of the wire, the portion of the wire below S is in equilibrium under the stress across S and the deflecting couple of moment M.

Hence the stress across any horizontal section reduces to a couple M with its axis along the wire.

Therefore the strain is everywhere the same, and the wire is twisted uniformly throughout its length.

The *total twist* of the wire is measured by the angle ψ, and if l is the length of the wire, $\frac{\psi}{l}$ is called the *twist*.

If the length of the wire is halved, the total twist is evidently halved too, and by Hooke's law if the couple is doubled the twist is doubled.

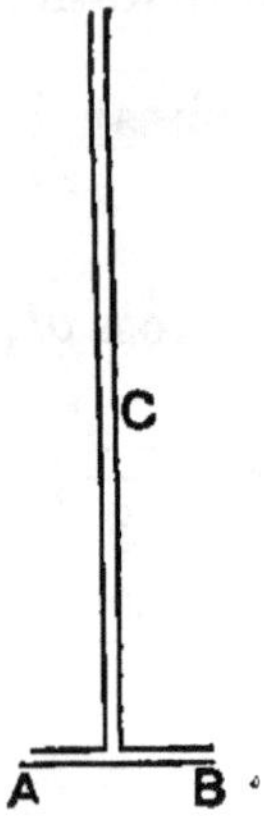

Hence the total twist is proportional to the deflecting couple and to the length of the wire.

Torsion is equivalent to shearing strain, for if the couple M produced any contraction, an opposite couple would produce expansion; the two couples however clearly produce similar strains, and therefore each can only produce shearing strain.

C

A B

Fig. 122.

It can be proved mathematically that if the cross-section of the wire is circular, it undergoes no deformation by torsion; but a square or elliptic section is distorted by twist and ceases to be plane.

It will be shown in Chapter X that the total twist produced by a given couple in a wire of given length and material is inversely proportional to the fourth power of the radius. Therefore if M is the twisting couple, l, r the length and radius of the wire, ψ the total twist,

$$M = \frac{Cr^4\psi}{l},$$

where C is a constant depending on the material of the wire, which is called the coefficient of torsion. It can be shown that $C = \frac{\pi n}{2}$.

We may determine C (or n) from the oscillations of a wire as follows.

Suspend from the wire a cylinder, the axis of which is the axis of the wire produced. Let H be the mass of the cylinder, K its radius of gyration about the axis.

The couple exerted on the cylinder by the wire, when the angular displacement is ψ, is $\frac{Cr^4\psi}{l}$, and HK^2a is the moment of the angular acceleration. Therefore

$$HK^2a = \frac{Cr^4\psi}{l}.$$

Hence the vibration is simply harmonic, and its period T is $2\pi\sqrt{\frac{HK^2l}{Cr^4}}$.

$$\text{Therefore} \qquad C = \frac{4\pi^2 HK^2 l}{r^4 T^2}.$$

The Torsion Balance.

When the rigidity, radius and length of the wire are known, the moment of any couple applied to the wire can be determined from observation of the twist produced; but the magnitudes of different couples may be compared by simple comparison of the corresponding twists.

A wire used in this manner, provided with proper arrangements for adjusting and measuring the twist, is called a Torsion Balance.

As an illustration of its use, we shall describe Cavendish's method of detecting and measuring the attraction (due to gravitation) between two bodies.

AB is a light metal rod of length $2a$, suspended by a long fine metal wire; at its ends it carries two small equal spheres A and B.

P and Q are two large spheres of lead with their centres in the same horizontal plane as those of A and B; they are attached to the ends of a lever and their centres are equidistant from the middle point of AB.

By a motion of the lever the centres of the lead spheres can be moved into the positions M and N, or M' and N', the lines AM, AM', being perpendicular to AB.

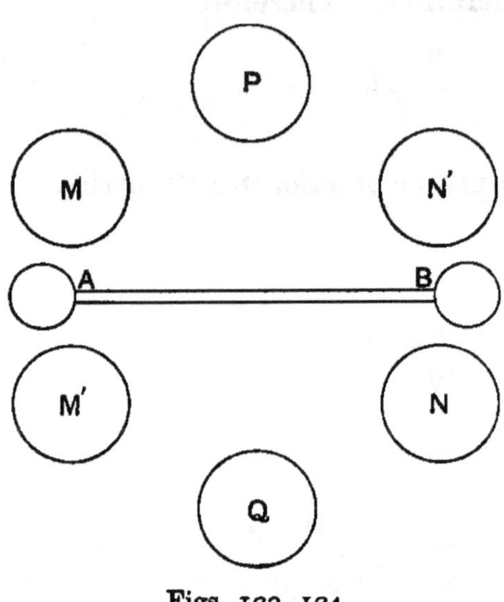

Figs. 123, 124.

Let m, m' be the masses of the spheres P and A; when they are at a distance x apart, the force between them is

$$\frac{kmm'}{x^2}.$$

When the lead spheres are at P and Q, they exert on the whole no couple on the spheres A and B; when they are at M and N they exert a couple $2aF$, where F is the force exerted by M on A.

Therefore if C is the coefficient of torsion, the angle θ through which the wire is twisted when the lead spheres are displaced from P, Q, to M, N, is given by

$$2aF = \frac{Cr^4\theta}{l}.$$

If the spheres are displaced from P, Q, to $M'N'$, the deflection is the same, but is in the opposite direction.

Therefore if we observe the angle ϕ through which AB turns when the spheres are moved from M, N, to M', N',

$$4aF = \frac{Cr^4\phi}{l}.$$

And since l, a, r can all be measured, F can be found.

To deduce the mean density d *of the Earth.*

Since $k = \frac{Fx^2}{mm'}$, k can be found by this experiment, and if M and R are the Earth's mass and radius respectively,

$$g = \frac{kM}{R^2},$$

$$\text{and} \quad d = \frac{M}{\frac{4}{3}\pi R^3} = \frac{3gR^2}{4\pi k}.$$

Thus M and d can both be found.

The best experiments made by this method give 5·48 as the Earth's mean density.

§ 14. Flexure.

The complete investigation of the behaviour of a rod under flexure requires advanced mathematical treatment; its principal results, however, are easy to express.

Consider a bar of oblong cross-section fixed at one end and bent by the application of a force F at the other, the force being parallel to one side of the cross-section. The portion of the bar between the free end and a cross-section S is in equilibrium under the force F and the stress across S, the weight being neglected. The stress across S therefore reduces to a force F and a couple of moment Fx, where x is the distance of S from the free end of the bar.

Hence when a bar is bent by a single force F the stress is not the same across all sections.

The Bending Moment at any section S is the moment, round a point in S, of all the forces between S and the end of the bar.

Uniform Flexure. Let the bar rest symmetrically on

supports A and B, and be bent by equal weights P applied to its ends C, D.

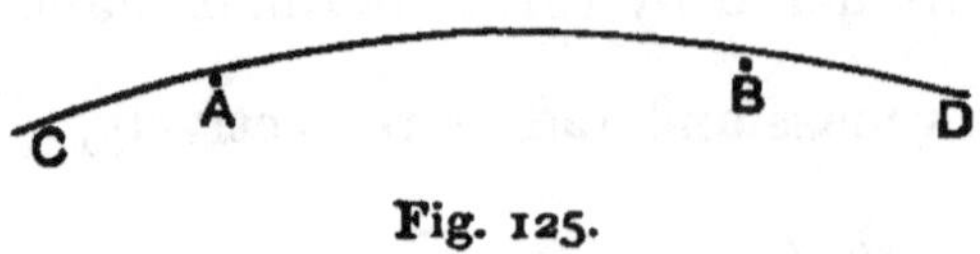

Fig. 125.

The stress across the section of the rod at the middle point must be entirely horizontal, for an upward force on one half of the rod would imply the existence of a downward force on the other, and considerations of symmetry render this impossible.

Hence the conditions for equilibrium of either half of the rod show that the resistances at A and B are equal to P.

If $AC = BD = l$, the Bending Moment at any point of AB is Pl, and since the bending moment is everywhere the same, the strain must be the same. Hence the bar when bent assumes a circular form.

The complete investigation gives the following results:—

The mean line, i.e. the line through the centres of the cross-sections, of the wire is bent into a circular form HK, and is not altered in length by strain.

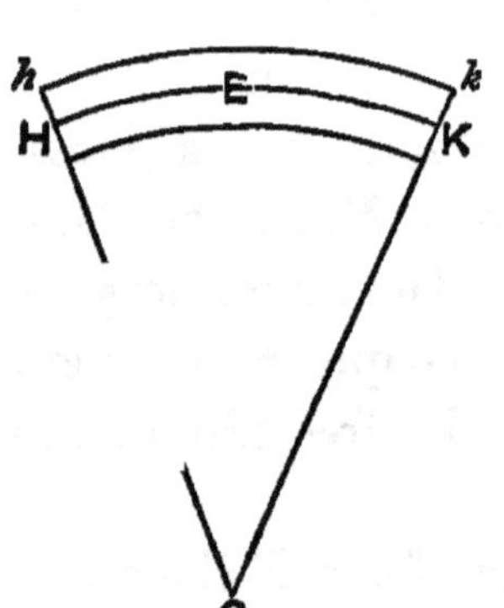

Fig. 126.

All planes, initially perpendicular to the mean line, become by strain planes through the point C (the centre of HK) perpendicular to HCK.

Lines parallel to the mean line become by strain circles with their centres on the perpendicular from C to the plane HCK.

If the whole rod is imagined to be divided parallel to its length into filaments of exceedingly small thickness, each filament shrinks or swells laterally with the same freedom as if it were separated from the rest of the substance.

Let hk be the axis of any filament in the plane HKC, y its initial distance from HK, regarded as positive when above HK, $CH = R$.

$$\text{Then } \frac{hk}{Ch} = \frac{HK}{CH} = \frac{hk - HK}{hH}.$$

$$\text{Therefore } \frac{hk - HK}{HK} = \frac{hH}{CH} = \frac{y}{R}, \text{ if } \frac{y^2}{R^2} \text{ is neglected.}$$

Therefore the stress across a section at distance y from HK is a tension $\frac{\mu y}{R}$. At points below HK the stress is a pressure.

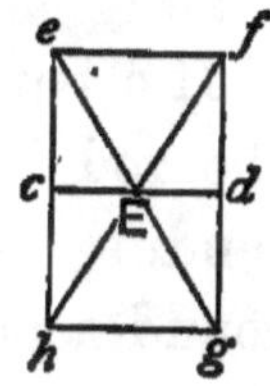

Fig. 127.

Let $efgh$ be a cross-section of the bar, E its centroid, $ef = a$, $fg = b$.

Imagine the bar to be divided into thin filaments by horizontal planes parallel to ef, the cross-sections of the filaments being of area $S_1, S_2, \ldots S_n$ respectively. Let $y_1, y_2, \ldots y_n$ be the distances of the axes of the filaments from E before strain.

The total stress on S_1 is $\frac{\mu y_1}{R} S_1$, and is a pressure or tension according as y_1 is positive or negative.

Its moment round the axis cd parallel to ef is $\frac{\mu y_1^2}{R} S_1$.

The total stress on $efgh$ is made up of the stresses on $S_1, S_2, \ldots S_n$, and reduces to a couple of moment

$$\frac{\mu}{R}\left(y_1^2 S_1 + y_2^2 S_2 + \ldots + y_n^2 S_n\right).$$

$$\text{But} \qquad y_1^2 S_1 + y_2^2 S_2 + \ldots + y_n^2 S_n = K^2 . ab,$$

where K is the radius of gyration of $efgh$ about cd.

Therefore if M is the bending moment $M = \frac{\mu I}{R}$, where I is the moment of inertia of the cross-section about the horizontal line through its centroid.

μI is called the flexural rigidity.

By the method of § 3, Chapter IV, it can be shown that $I = \frac{ab^3}{12}$.

If the rod is turned through a right angle so as to bend about an axis parallel to fg, the flexural rigidity is $\frac{\mu a^3 b}{12}$ and the flexural rigidities in the two cases are as $b^2 : a^2$.

If a bending moment round any other axis is applied to the rod its forces must be resolved into components, one tending to produce flexure round cd, the other, flexure round a perpendicular to cd. As these flexures are not in the same proportion as the forces which produce them, the relation of the bending couple to the resultant flexure is in this case somewhat more complex.

Work done by Stress.

Hydrostatic Stress. Let the volume of a body be V when free from stress (or subject to negligible stress), $V - v$ when subject to hydrostatic pressure P.

In estimating the work done we suppose that no kinetic energy is generated, or that the stress is increased very gradually; we also assume that Hooke's law is satisfied.

The strain consists only in change of volume, and no change of shape occurs; hence the result obtained is the same as for a body of no rigidity, and we may give the body any form we please.

Imagine it to be contained in a cylinder of section S,

fitted with a piston which moves in through a distance d, when the pressure is increased to P.

When the volume is $V - nv$, the pressure is nP, n being a fraction, and the distance through which the piston has been displaced is nd.

Since the total pressure on the piston is $nP \times S$, the force on the piston is proportional to the displacement, and (Chapter III, § 3) the work done in the displacement d is $\frac{1}{2} PS . d$.

Now $Sd = v$.

Therefore the work done is $\frac{1}{2} Pv$.

If the pressure on the piston is originally P_0, the pressure when the volume has been diminished by nv will be $P_0 + nP$, and the total work done is $P_0 v + \frac{1}{2} Pv$.

If the initial and final pressures are P_0, P_1 the work done is

$$\frac{1}{2}(P_0 + P_1)\, v.$$

Similarly if a wire or string increases in length by l under a stretching force W, the work done in strain is $\frac{1}{2} Wl$.

Expansion of a gas.

The volume of a gas varies considerably, not only with the pressure, but also with the temperature; Hooke's law is not satisfied except for very small changes of volume.

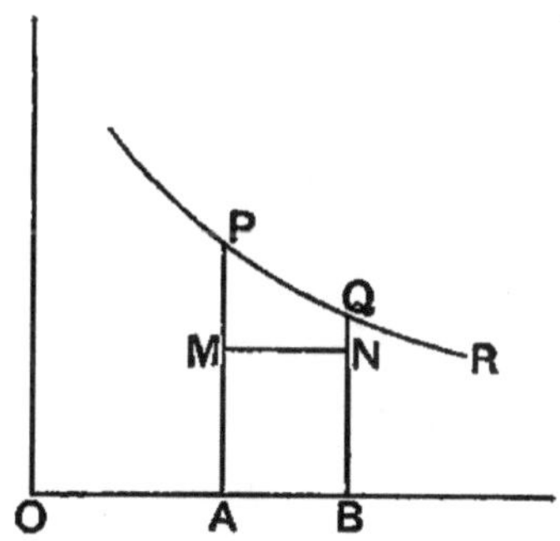

Fig. 128.

Let a gas contained in a cylinder expand from volume v_0 to v, its temperature being so regulated that the pressure is constant and equal to p. The work done by the gas in expanding is $p(v-v_0)$, and if $OA = v_0$, $OB = v$, $AM = p$, the rectangle $AMNB$ represents the work done by the gas in expanding.

Now let the pressure and volume of a given mass of gas vary in any manner, and let the abscissa and ordinate of a point represent the volume and pressure of the gas at a given instant. Then a continuous curve PQR represents the states through which the gas passes in expanding, and the argument of Chapter III, § 3, shows that the work done in expansion from v_0 to v is represented by the area $APQB$.

CHAPTER VIII.

HYDROSTATICS.

§ 1. IN a fluid at rest the only possible stress is hydrostatic, and, except under unusual conditions, this stress is a pressure.

Fluids are divided into two classes, liquids and gases.

Every liquid can be brought into the gaseous state by a sufficient increase of its temperature, and every gas can be liquefied by cooling it and increasing to a greater or a less extent the pressure to which it is subjected.

By suitable processes a liquid can be continuously and imperceptibly changed into a gas, but in the case of all known substances, under ordinary atmospheric pressure and at ordinary temperatures, there is a marked distinction between liquids and gases.

A liquid is only very slightly compressible, and may for our purposes be considered as incompressible. The density of a liquid will therefore be regarded as constant.

A gas on the other hand is easily compressed. In the case of air and other gases which are not readily brought to the liquid state, the density is proportional to the pressure, when the temperature is kept constant.

Thus if external pressure is removed the gas expands indefinitely.

§ 2. **In a heavy fluid at rest, the pressure is the same at all points which are in the same horizontal plane.**

Consider a cylindrical portion of fluid with horizontal axis, terminated by parallel vertical planes, and let A be the area of either plane end of the cylinder, p the pressure at one end, p' the pressure at the other end.

Fig. 129.

The cylinder is in equilibrium under its own weight, the resultant of the pressures on the curved surfaces, and the total pressures pA, $p'A$ applied to the ends of the cylinder.

The only forces in the direction of the axis of the cylinder are pA, $p'A$.

Therefore these are equal and opposite or $p = p'$.

Cor. Since the cylinder has no vertical motion, the resultant of the pressures on the curved surface of the cylinder passes through the centre of mass of the cylinder, is directed vertically upwards, and is equal in magnitude to the weight of the cylinder.

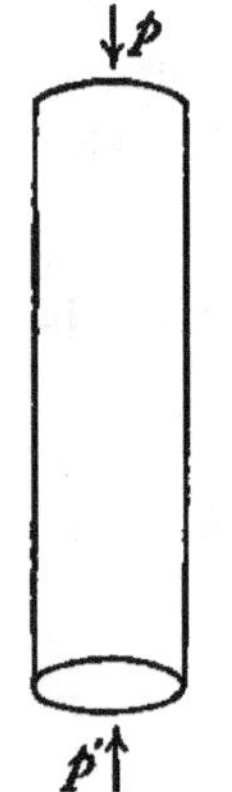

Fig. 130.

§ 3. **To compare the pressures at different depths in the fluid.**

Let the axis of the fluid cylinder be vertical, the pressures on the curved surface are then horizontal, and the only vertical forces are pA and W (the weight of the cylinder), downwards, and $p'A$ upwards.

Therefore $$p'A = pA + W.$$

If the density of the fluid is uniform and equal to ρ, and h is the height of the cylinder,

$$W = \rho g h A \quad \text{and} \quad p' = p + \rho g h.$$

If π is the pressure of the atmosphere the pressure at a depth h in the liquid is $\pi + \rho g h$.

The common surface of two liquids of different density which do not mix is a horizontal plane.

For let AB, PQ be horizontal lines in the same vertical plane AQ, AB lying in the upper liquid, and PQ in the lower.

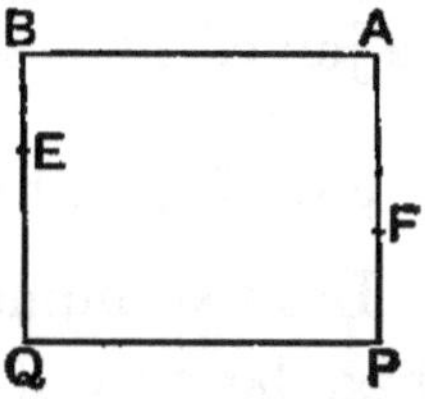

Then it has been proved that the pressures at A and B are equal, and that the pressures at P and Q are equal.

Fig. 131.

Let p, p' be the pressures at A and P; E, F the points where BQ, AP meet the common surface of the liquids, ρ, ρ' the densities of the upper and lower liquid respectively.

Then if $BQ = h$, $BE = x$, $AF = y$,

$$p' - p = g\{\rho x + \rho'(h - x)\} = g\{\rho y + \rho'(h - y)\}.$$

Therefore $x = y$ unless $\rho = \rho'$, and therefore E and F are in the same horizontal plane.

Also the equilibrium is not stable (i. e. cannot exist) unless the lower liquid is the denser.

Hence also the free surface of a liquid is horizontal, for ρ may denote the mean density of air in the neighbourhood of the free surface.

In the above propositions it is assumed that g is the same in magnitude and direction at all points of the liquid. This ceases to be justifiable when a large liquid surface, as that of a lake or ocean, is considered.

Cor. Let a vessel containing liquid move vertically upward with acceleration a.

Considering as before the motion of a portion of liquid bounded by a cylindrical surface with vertical axis, we have $A(p' - p - \rho gh)$ as the resulting force upwards and ρhA is the moving mass.

$$\therefore \quad \rho ha = p' - p - \rho gh,$$

$$\text{or} \quad p' = p + \rho h(g + a).$$

When liquid is contained in a curved vessel of any form, the laws of distribution of pressure that have been laid down are still true, for if two points in the vessel cannot be joined directly by a vertical or horizontal line they can be joined by a series of such lines, all lying in the liquid.

§ 4. Equilibrium of two liquids in a U-tube.

Let two liquids such as mercury and water, which do not mix, be placed in a U-tube, and let P, Q be the free surfaces of the liquids, S their common surface of intersection.

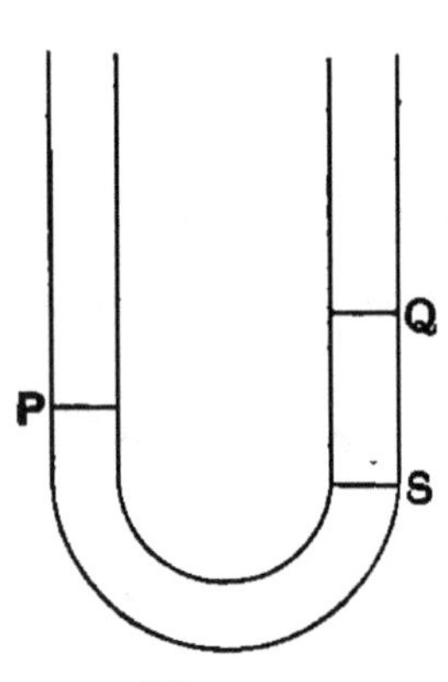

Fig. 132.

Let h, h' be the vertical heights of P and Q above S, ρ the density of the liquid between P and S, ρ' the density of the other liquid, π the atmospheric pressure.

Then $\pi + \rho g h$ is the pressure at S due to the liquid PS, and $\pi + \rho' g h'$ is the pressure due to the liquid QS.

Therefore $\pi + \rho g h = \pi + \rho' g h'$.

And $\rho h = \rho' h'$.

Thus the heights to which the liquids rise above S are inversely as their densities.

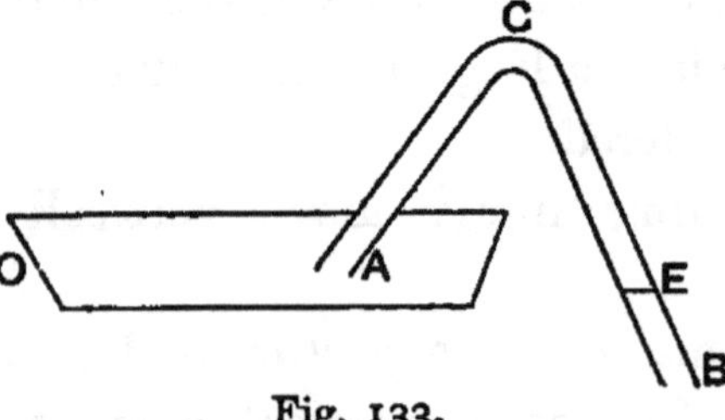

Fig. 133.

The Siphon. The siphon is a bent tube ACB employed to transfer liquid from one vessel O to another at a lower level.

Let the arm CA be immersed in a liquid of density ρ, which fills the portion ACE of the tube.

Then if the vertical depth of E below the free surface in

O is r, the pressure of the liquid at E is $\pi + \rho g r$, π being the atmospheric pressure. And the opposing pressure at E is only π.

Hence the liquid flows out at E, at a rate which increases with r.

In order that the siphon may work, it is necessary that it should first be filled with so much liquid that the free surface in the arm BC (at E) lies below the free surface in the vessel from which water is drawn.

§ 5. Archimedes' principle.

Consider any portion A of a fluid in equilibrium, under its own weight and the pressures of the surrounding fluid on its surface. The resultant of the pressures is a single force acting upwards through the centre of mass and equal to the weight of the fluid in A.

If a solid body occupies the space A, the pressures at its surface are the same as when A was occupied by fluid.

Therefore the resultant pressure of a fluid on a body is equal to the weight of the fluid which occupies the same volume as the immersed portion of the body (called for brevity the fluid displaced); and the resultant acts vertically upwards through the centre of mass of the displaced fluid. This is Archimedes' principle.

Let the body be a homogeneous solid completely immersed in a homogeneous liquid. Then the centres of mass of the solid and the displaced liquid coincide. And if V, ρ be the volume and density of the body, σ the density of the liquid, $V(\rho - \sigma)g$ is the resultant downward force on the body.

Thus a body when wholly immersed sinks or rises according as its density is greater or less than that of the liquid.

Let the body be only partially immersed, displacing a volume of liquid v.

Then $(V\rho - v\sigma)g$ is the downward force, and the body floats in equilibrium if $V\rho = v\sigma$, and if the centres of mass of the body and the liquid displaced are in the same vertical line.

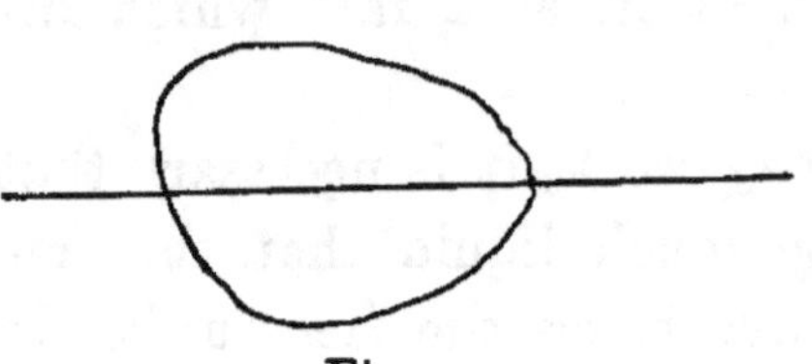

Fig. 134.

Let a body float in two liquids (Fig. 134) so that its surface meets their common horizontal surface.

This is the case of a body floating in water, when the pressure of the atmosphere is taken into account.

Let σ, σ' be the densities of the upper and lower liquids, v, v' the volumes displaced; then $v + v'$ is the volume of the body, and if ρ is its density there is equilibrium when

$$v\sigma + v'\sigma' = \rho(v + v').$$

If the immersed body is hollow, it will float in a liquid of less density than its own; hence the possibility of using iron ships.

Experimental proof of Archimedes' principle.

From one pan of a balance two cylinders are hung, one below the other. The upper cylinder is hollow, its internal volume being exactly the volume of the lower cylinder. The beam of the balance is rendered horizontal by placing masses in the other pan.

A beaker of water is now brought under the two cylinders so that the lower cylinder is completely immersed. The upward pressure of the water on this cylinder destroys equilibrium, which is restored if the upper cylinder is filled with water. Hence the principle is experimentally demonstrated, for the water in the upper cylinder occupies the same volume as the lower cylinder.

The best evidence of the truth of Archimedes' principle is afforded by the fact that the same value of the density of a body is obtained, whether it is determined by the hydrostatic method, or by direct measurement of the mass and volume of the body.

§ 6. Methods of finding the density of a body.

(1) The Hydrostatic Method.

Suspend the body from the right-hand pan of the balance by a hair, and place a mass M in the left-hand pan, rendering the beam horizontal. We suppose that the balance is accurate, and that the weight of the air can be neglected.

Now immerse the body, still hanging by the hair, in a beaker of water; the tension in the hair is then diminished by the weight of water displaced by the body, and a mass m must be withdrawn from the left-hand pan to restore the beam to the horizontal position.

Then m is the mass of water displaced, and if σ is the density of water, $\frac{m}{\sigma}$ is the volume of the body, and $\frac{M\sigma}{m}$ is its density.

The density of a liquid can also be determined in the same way; for if the beaker contains some other liquid of density σ', and m' is the mass of this liquid displaced by the body, the volume of the body is $\frac{m'}{\sigma'}$.

$$\text{And } \frac{m'}{\sigma'} = \frac{m}{\sigma} \text{ or } \sigma' = \frac{m'\sigma}{m}.$$

If the volume V of the suspended body is known, $\sigma' = \frac{m'}{V}$, and the density of the liquid can be deduced from this formula, without assuming the density of any other body to be known.

The density of water at different temperatures has been determined by weighing a carefully turned glass cylinder of measured dimensions in water, and in air; the mass of water displaced by the cylinder was thus ascertained.

(2) The Specific Gravity Bottle.

This is a small stoppered flask of about 30 c.cm. content, with a narrow neck, round which a circular mark is drawn.

To determine the density of a liquid by the flask, dry the flask and determine its mass when empty; then fill it up to the mark with liquid, and find the increase of mass. Let this be M_1.

Empty and clean the bottle, and fill it up to the mark with water. Let the mass of water determined by the balance be M_2.

Then if σ' is the density of liquid, σ that of water,

$$\sigma' = \frac{M_1 \sigma}{M_2}.$$

Reduction to a vacuum.

In accurate experimental work, the above results require a correction for the air displaced.

Let ρ be the density of a body, V its volume, a and σ the densities of air and water.

Then the mass determined by weighing in air is $V(\rho - a)$, and determined by weighing in water is $V(\rho - \sigma)$.

Therefore in (1) $V(\rho - a) = M$.

And $V(\rho - \sigma) = M - m$.

Whence if a and σ are known, ρ and V can be found.

Again, in (2) let N be the true mass of liquid, V the volume of the flask up to the mark.

Then $$N - Va = M_1,$$

or $$V(\sigma' - a) = M_1.$$

Similarly $$V(\sigma - a) = M_2.$$

Whence $$\frac{\sigma' - a}{\sigma - a} = \frac{M_1}{M_2}.$$

And σ' can be found if σ and a are known.

(3) Hydrometers.

The Hydrometer is a float of glass or metal, which is used to determine the density of liquids, and sometimes of solids.

Hydrometers are divided into two classes, hydrometers of constant immersion and variable load, and hydrometers of variable immersion and constant load.

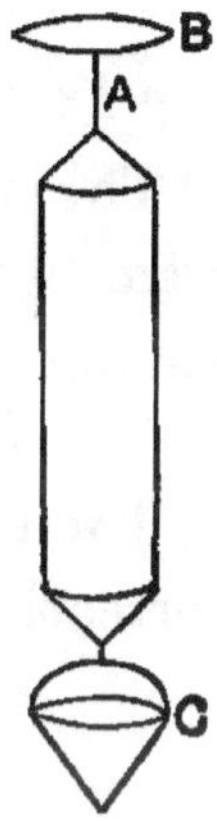

Nicholson's hydrometer is represented in the figure; it is made of brass and loaded at the bottom, so that it floats in stable equilibrium, with its axis upwards.

Fig. 135.

B is a platform on which weights can be placed, which sink the hydrometer till the mark A, on the wire supporting B, is in the surface of the liquid.

Let P be the mass of the hydrometer, M the mass on the platform B when the hydrometer is immersed in water to the mark A, N the mass when the hydrometer is similarly immersed in some other liquid of density σ'.

Then the masses of liquid displaced are $P+M$, $P+N$ respectively, and their volumes are the same.

Then $$\sigma' = \frac{P+N}{P+M}\sigma.$$

The hydrometer can also be used to find the density of a solid as follows.

Let a mass M sink the hydrometer to the mark A.

Place the solid on the platform B, with masses M_1, which sink the hydrometer to the mark.

Place the solid on the platform C below the liquid surface, and place masses M_2 on B, till the hydrometer sinks to the mark.

Let V be the volume, and ρ the density of the body. Then if we neglect the correction for the air displaced,

$$V\rho = M - M_1 \quad \text{and} \quad V\sigma = M_2 - M_1;$$

$$\therefore \quad \rho = \frac{M - M_1}{M_2 - M_1}\sigma.$$

Hydrometers of variable immersion.

The upper part of these instruments is a vertical glass stem, which is graduated to indicate in arbitrary units the volume immersed.

Let V_1, V_2 be the volumes immersed in different liquids.

Then if σ_1, σ_2 are the densities of the liquids, and the correction for air is neglected

$$V_1\sigma_1 = V_2\sigma_2.$$

And thence the densities can be compared.

The Specific Gravity of a body is the ratio of its density to that of a standard substance, usually water at its maximum density.

In the C. G. S. system of units σ is very nearly 1, and there is no sensible distinction between the density and the specific gravity of a body; but when the foot and pound are units, the densities of most bodies are represented by inconveniently high numbers, and simplicity is gained by referring to specific gravity rather than to density.

§ 7. Total pressure on a plane surface in a liquid. Centre of Pressure.

Let S be the surface, ρ the density of the liquid.

Divide the surface into small elements $S_1, S_2, \ldots S_n$ at depths $z_1, z_2, \ldots z_n$ below the liquid.

Then, if there is no pressure on the free surface, the pressure on the element S_1 is $g\rho z_1$, and the sum of the total pressures is

$$g\rho(z_1 S_1 + z_2 S_2 + \ldots + z_n S_n).$$

But if z is the depth of the centroid of S,

$$zS = z_1 S_1 + z_2 S_2 + \ldots + z_n S_n.$$

And the total pressure is $g\rho z S$.

If the pressure on the free surface of the liquid is π, the total pressure is $(\pi + g\rho z) S$.

Since the pressures on S are similarly directed parallel forces, they have a single resultant, and the point when the line of action of this resultant meets S is called the centre of pressure.

To find the centre of pressure.

Through each point of the boundary of S draw a vertical line; these lines form a prism or cylinder and meet the liquid surface in B.

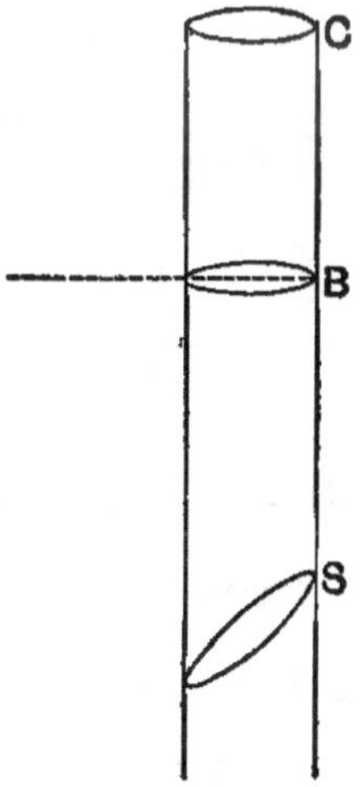

Then the solid bounded by B, by S, and by the cylinder, is in equilibrium, and the vertical component of the total pressure on S must be equal to the weight of the solid.

Therefore the centre of pressure is the point where the vertical line through the centroid of the solid meets S.

Fig. 136.

If the pressure at the free surface of the liquid is π, produce the cylinder upwards from B through a distance h, such that $\pi = \rho g h$.

Then if C is the horizontal plane in which the cylinder

terminates, the centre of pressure lies in the vertical line through the centroid of the solid bounded by C, by S, and by the cylinder.

Example.—A rectangle is immersed in liquid, with one side in the free surface and the opposite side at a known depth.

To find the total pressure and the centre of pressure.

Let the plane of the figure AON be a vertical plane through the centroid E of the rectangle, cutting the liquid surface in ON, and the rectangle in OA.

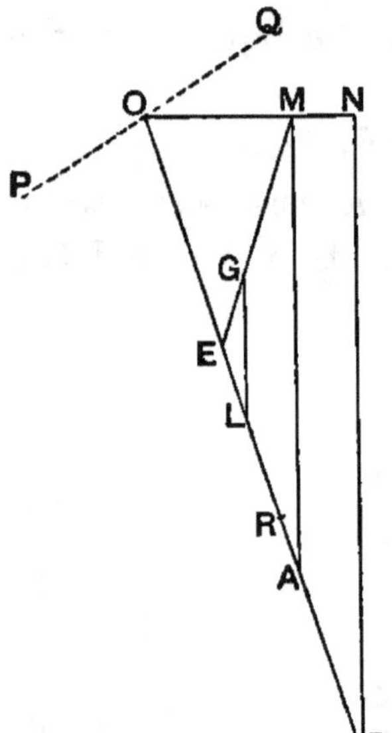

Fig. 137.

Draw the vertical AM; find the centroid G of the triangle OMA, and draw GL parallel to AM, and MG meeting OA in E.

Let h be the length of PQ, the horizontal side of the rectangle.

Then if $AM = l$, and $OA = k$, the depth of E is $\frac{1}{2}l$, and the surface of the rectangle is hk.

Therefore $\frac{1}{2}hkl\rho g$ is the total pressure required.

Again, L is the centre of pressure, and

$$\frac{AL}{LE} = \frac{MG}{GE} = 2.$$

Therefore $AL = \frac{1}{3}AO$, or $OL = \frac{2}{3}OA$.

This is true for all inclinations of the rectangle to the horizon.

If OA is produced to B, so that $OB = k_1$, and BN is drawn parallel to AM, $BN = \frac{k_1 l}{k}$.

Hence the total pressure on the rectangle contained by OB and PQ is $\frac{1}{2}\frac{hk_1^2 l\rho g}{k}$, and the centre of pressure is at R, when $OR = \frac{2}{3}OB$.

Therefore the total pressure on the rectangle contained by AB and a parallel to PQ is $\frac{1}{2}hl\rho g\frac{k_1^2 - k^2}{k}$, and the centre of pressure is at a point V in OB, such that

$$OV.(k_1^2 - k^2) = \tfrac{2}{3}(k_1^3 - k^3),$$

$$\text{or} \quad OV = \tfrac{2}{3}\frac{k_1^2 + k_1k_2 + k_2^2}{k_1 + k_2}.$$

This result might have been obtained directly by finding the centroid of the area *AMNB*.

The above method of finding the centre of pressure fails when the immersed surface is vertical.

This case may be treated as follows:—

Let *C* be a cylinder in the liquid with its axis horizontal, bounded by a vertical plane *A* perpendicular to the axis, and a plane *T* which intersects *A* in a horizontal line.

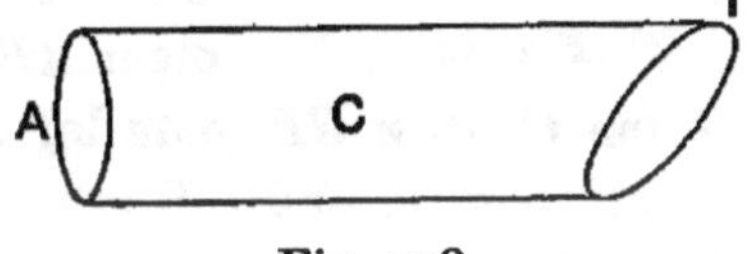

Fig. 138.

Since this portion of the liquid is in equilibrium the resultant pressure on *A* must be equal and opposite to the resultant horizontal pressure on *T*, and the centres of pressure of *A* and *T* are in the same horizontal line, parallel to the axis of the cylinder.

Thus, let *ABCD* be a vertical rectangle, *AB* being horizontal; draw *AE*, *BF* perpendicular to the plane *ABCD*, meeting any plane through *CD* in *E* and *F*.

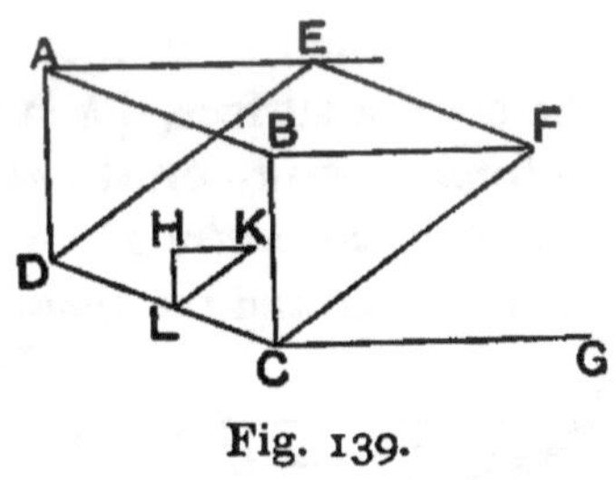

Fig. 139.

The centres of the pressure of the rectangles *CDEF*, *ABCD* are in the same horizontal line parallel to *AE*.

Bisect *CD* in *L*, and draw *LH*, *LK* parallel to *AD*, *DE* respectively; make $LK = \frac{1}{3}DE$, and draw *KH* parallel to *AE*. Then *H* is the centre of pressure of *ABCD*.

$$\text{But} \qquad \frac{LH}{LK} = \frac{AD}{DE}.$$

$$\text{Therefore} \qquad LH = \tfrac{1}{3}AD.$$

Thus the results obtained before are true when the rectangle is vertical.

To find the centre of pressure of a triangle with one side in the liquid surface.

Let ABC be the triangle, AB being in the liquid surface. From C draw CD vertically upwards to meet the surface in D.

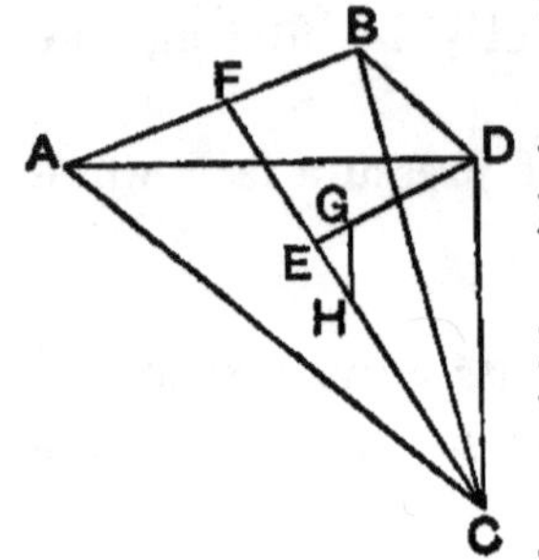

Fig. 140.

Then there is no pressure on the face ABD of the tetrahedron $ABCD$, and the pressures on CDA and CDB are horizontal.

Therefore the vertical component of the pressure on ABC is equal to the weight of the tetrahedron.

Let E be the centroid of ABC, join CE and DE, and make $DG = \frac{3}{4} DE$. Produce CE to meet AB in F.

From G draw GH parallel to CD. Then G is the centroid of the tetrahedron and GH is vertical.

Therefore H is the centre of pressure of the triangle.

And $$\frac{CH}{CE} = \frac{DG}{DE} = \frac{3}{4}.$$

Therefore $$CH = \tfrac{3}{4} CE = \tfrac{1}{2} CF.$$

Thus if Δ is the area of the triangle, h the depth of the vertex C below the surface, $\frac{1}{2} h$ is the depth of the centre of pressure, and the total pressure on the triangle is $\frac{1}{3} h \Delta \rho g$.

To find the centre of pressure of a triangle with one vertex in the liquid surface and the opposite side horizontal.

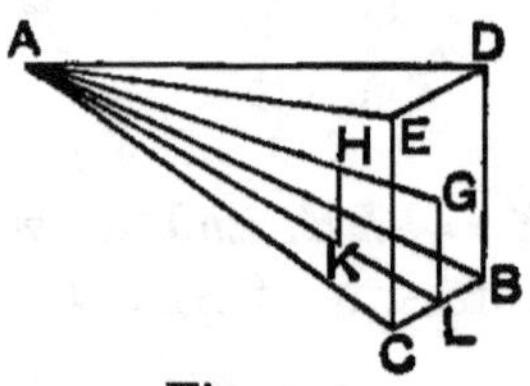

Fig. 141.

Let ABC be a triangle, having the vertex A in the surface of the liquid, and the base BC horizontal at a depth h.

Draw BD, CE vertically upwards meeting the liquid surface in D and E. Join AE, AD, DE.

Then the vertical pressure on ABC is the weight of the pyramid $ADECB$ described on the rectangular base $BDEC$.

This is twice the tetrahedron $AEBC$, and therefore the vertical pressure is $\frac{2}{3} h \Delta \rho g$.

Let G be the centroid of the rectangle $BDEC$. Then if a point

H is taken on AG such that $AH = \frac{3}{4} AG$, H is the centroid of the pyramid.

If HK and GL are drawn paralled to the vertical CE, meeting ABC in K and L, $CL = LB$, and $\frac{AK}{AL} = \frac{AH}{AG} = \frac{3}{4}$.

Thus the position of K is known.

Pressure on a curved surface.

The total pressure on a curved surface is the sum of the total pressures on each element of it. As in the case of a plane surface it is $p\,S$, where S is the area of the surface, p the pressure at its centroid.

In some cases the total pressure on a curved surface is equivalent to a single force, which is called the resultant pressure.

When the curved surface is bounded by a plane curve the resultant pressure may often be found from the conditions of equilibrium of the liquid enclosed between the surface and the plane through the bounding curve. If the weight of the liquid can be neglected, the resultant pressure on the curved surface is equal to that on the plane surface, which has the same boundary.

Example.—*A hemisphere, of radius r, is immersed in liquid with its plane face vertical; to find the resultant pressure on the curved surface.*

The horizontal component of the resultant pressure is equal to the total pressure on the plane surface, and is $\pi r^2 p$, if p is the pressure at the centre of the sphere.

The vertical component of the resultant pressure is equal to the weight of the liquid displaced by the hemisphere.

§ 8. Application of the Principle of Work.

Let E be an enclosure filled with liquid at rest, C and D tubes opening into E and closed by pistons A and B.

Let a and b be the areas of the faces of the pistons, p and q the pressures on the pistons when there is equilibrium.

Then in any small displacement under these pressures the work done is zero.

First, let no external forces act on the mass of liquid.

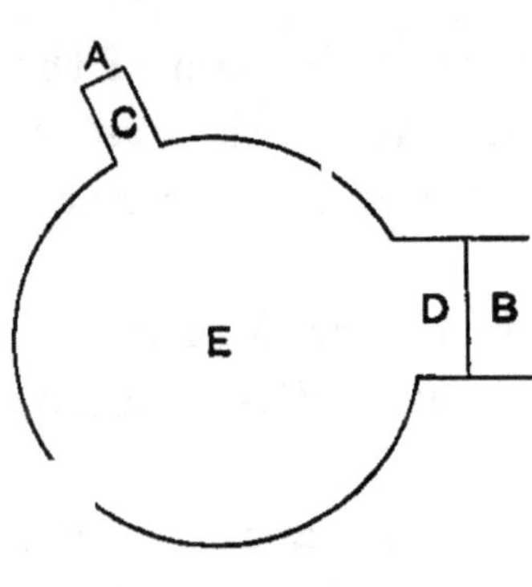

Fig. 142.

The pressures at the fixed surface of the enclosure do no work, as they are everywhere perpendicular to the displacement.

If A is displaced inwards through a distance d, then B must be displaced outwards through a distance e, and since the liquid is incompressible

$$ad = be.$$

The forces exerted by A and B on the liquid are pa and qb respectively, and the work done by them is $pad - qbe$.

But this must be zero and $ad = be$.

Therefore $p = q$, or the pressure on the two pistons is the same.

Thus, assuming that the liquid is incompressible and that the pressure on any surface is perpendicular to it, it follows from the Principle of Work that the pressure throughout the liquid is uniform when no forces act on the mass of liquid.

Secondly, let forces which have a potential act on the liquid.

If V_1 be the potential at A, V_2 the potential at B, $V_1 - V_2$ is the work done in displacing unit mass from A to B.

Now the displacement already considered is equivalent

to the displacement of a mass ρad from A to B, ρ being the density of the liquid. Therefore since there is equilibrium

$$pad + \rho ad (V_1 - V_2) = qad,$$

whence $$p + \rho V_1 = q + \rho V_2. \qquad (1)$$

Since $p = q$, when $V_1 = V_2$, the surfaces of equal pressure are the equipotential surfaces.

Example.—A liquid sphere of radius a is at rest under the mutual gravitation of its parts; to determine the pressure at a point P, at distance x from the centre.

By Chap. VI, § 15, the force which acts on unit mass at P is $\frac{4}{3}\pi k\rho x$. Hence (Chap. III, § 4) the difference of the potentials at the surface and at P is $\frac{2}{3}\pi k\rho\,(a^2 - x^2)$.

Therefore the pressure at P is $\frac{2}{3}\pi k\rho^2\,(a^2 - x^2)$.

If the pressure at A is increased to $p + \pi$, it follows that the pressure at B becomes $q + \pi$.

Thus an increase of pressure at any point of a liquid at rest involves an equal increase of pressure at every other point.

This result is known as Pascal's principle of the transmission of fluid pressure.

§ 9. Bramah's Press.

If the section of the piston A is small compared with that of B, a very small force applied to A will balance a large force at B.

This fact has been employed in the construction of Bramah's Hydraulic Press, which in principle consists of a U-tube, with one branch much wider than the other, and a force pump with which a high pressure is created in the narrow branch of the tube.

The figure is a rough representation of the press.

A is a closed reservoir containing water.

B is the barrel of the pump communicating with *A* through the valve *C*, which is open or shut according as the pressure in *A* exceeds or falls short of that in *B*.

D is the piston of the force pump; it is raised or lowered by the lever *GEH*, working on a fulcrum at *G*, the hand at *H* supplying the power.

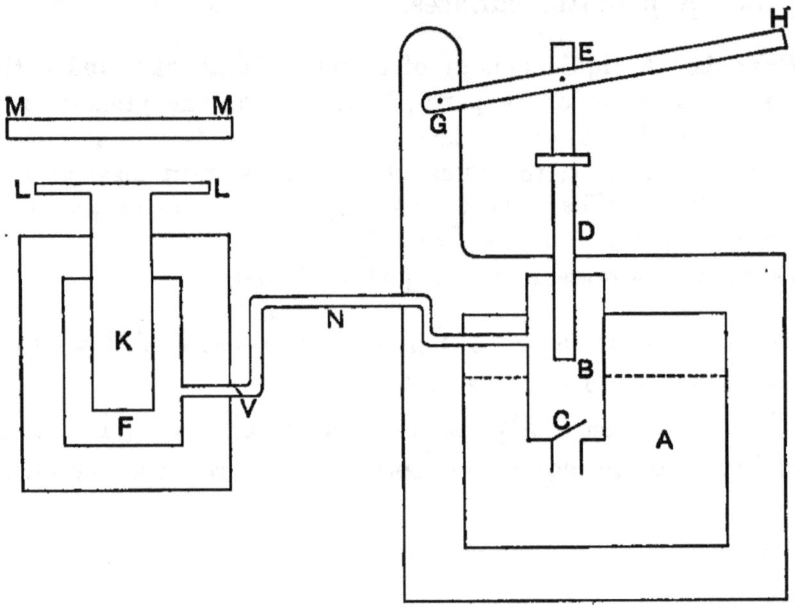

Fig. 143.

N is a pipe from *B* through which water can be admitted to the receiver *F*, passing on the way through a valve *V*, which opens towards *F*; and *V* is closed when the pressure in *F* exceeds that in *N*.

K is a piston working in the receiver *F* and terminating at the top in the stand tube. The body to be pressed rests on this stand, and is compressed between it and the horizontal plate *MM*.

When the piston of the pump is depressed, the valve *C* closes, and water is forced through *V* into *F* until the piston is in its lowest position. When the piston rises, the pressure in B diminishes and the valve V closes. The valve C opens and admits more water from A into the barrel B, until the piston reaches its highest position. The piston then descends again, performing the cycle just described.

We shall not consider resistances due to friction.

If P is the force exerted at H to keep the piston descending uniformly, $P \cdot \frac{GE}{GH}$ is the force acting at E.

Let S, s be the areas of the faces of the pistons D and K. The pressure on D is then $\frac{P}{s} \cdot \frac{GE}{GH}$, and this is transmitted to the receiver F. The horizontal pressures on the piston K have no effect, and the vertical *total* pressure on the piston is $P \cdot \frac{S}{s} \cdot \frac{GE}{GH}$, which is applied to the compression of the body resting on LL.

Making $\frac{S}{s}$ and $\frac{GE}{GH}$ large, a moderate force exerted at H brings a very large force into play at LL. Hence Bramah's Press is a machine of very great mechanical advantage.

The following difficulty may perhaps present itself to the reader.

A is the base of a conical beaker of area A, filled with liquid to a depth h. The total pressure on the base is ρghA.

But the base obviously supports the whole mass of liquid and the weight of this is greater than ρghA.

This is explained as follows—

Imagine a vertical cylinder described in the liquid on the base A.

The liquid outside this cylinder is kept in equilibrium by the horizontal pressures due to the cylinder, its own weight, W, and the pressures on the sides of the beaker. To these latter forces there is an equal and opposite action of the liquid on the sides of the beaker, and to balance this the base must exert an action on the sides whose vertical component is equal to the weight W, and is directed upwards.

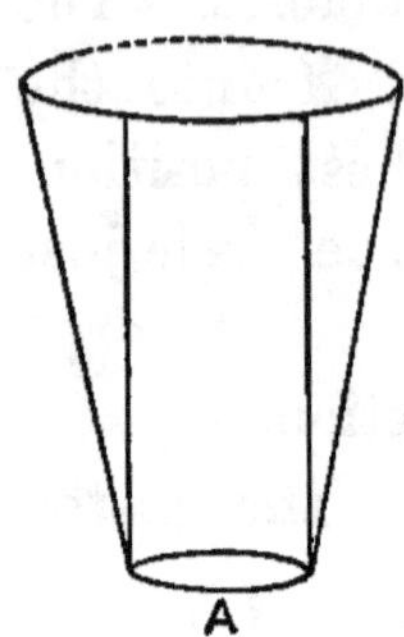

Fig. 144.

§ 10. Gases.

Gases are remarkable for their low density, their high compressibility, and their great expansion when heated.

All gases have weight; this can be demonstrated for air by exhausting a flask, weighing it, admitting air and then weighing again; the increase in weight is the weight of the air in the flask. If the volume of the flask is known, the density of the air can be determined in this way.

The density of dry air at 0° C under the pressure of a column of mercury 76 cm. high is ·001293. It can be demonstrated in like manner that other gases have weight and their densities can be found.

Atmospheric pressure. The following experiment was first performed by Torricelli.

A glass tube, rather more than 30 inches long and closed at one end, is filled with mercury, the other end temporarily closed with the thumb, and the tube then inverted in a mercury bath.

It will be found, on removing the thumb from the submerged end, that the mercury does not all escape from the tube, but that a column about 30 inches high remains in the tube.

Since the pressure at all points of the same liquid in the same horizontal plane is the same, the mercury in the free surface of the bath must be subject to a pressure equal to that exerted by this column.

This pressure was attributed by Pascal to the weight of the atmosphere. To test this idea, he carried the apparatus up the mountain Puy-de-Dôme in 1648, and found that the height of the column steadily diminished as greater altitudes were attained.

This result is now so well known, that the heights of mountains can be approximately estimated from the heights of the barometer column at their summits.

Measurement of atmospheric pressure.

If a Torricellian tube inverted in a mercury cistern be fitted with a vertical scale, by means of which the height of the column can be read, it forms a barometer, i.e. an instrument for measuring the pressure of the atmosphere; for if h be the height of the barometer column, and ρ the density of mercury, the pressure is ρgh.

The standard pressure (commonly called an atmosphere) is the pressure when the height of the barometer is 76 centimetres, and the temperature 0° C.

As the density of mercury under these conditions is 13·596, and $g = 981$, this is rather more than 10^6 dynes per sq. cm. A pressure of 10^6 dynes per sq. cm. might very well be made the scientific unit of pressure.

As the height of the barometer varies, mercury passes between the tube and the cistern, and the level of mercury in the cistern varies also. Hence to determine the length of the barometer column it is necessary to know the position of the free surface in the cistern as well as in the tube. This difficulty is best overcome in Fortin's barometer. Here the bottom of the mercury cistern is a bag of chamois leather, and a vertical screw working in a fixed nut presses against the bottom of the bag, so that, by raising or lowering the screw, the level of the mercury in the cistern is also raised or lowered. The tip of a fixed ivory pin pointing downwards indicates the zero of the barometer scale, and if the mercury surface is clean it can, by turning the screw, be adjusted very accurately to touch the tip of the pin.

Since the density of the air under standard conditions is ·001293, the height of the atmosphere if homogeneous would be $\frac{13\cdot596}{\cdot001293} \times 76$ cm., or about 5 miles.

Law of decrease of the pressure of the atmosphere.

The pressure of the atmosphere diminishes as the height above sea-level increases. Assuming that the atmosphere is at rest, and that its temperature is constant, we can find the law of variation of the pressure.

Consider a portion of the atmosphere contained between two horizontal planes at a small distance t apart; then if p is the pressure at the lower plane, p_1 the pressure at the upper plane, and ρ the average density of the air between the planes, $p = p_1 + \rho g t$.

Since t is very small, p_1 and p are very nearly equal, and, applying Boyle's Law (§ 11), we may write $\rho = kp$.

Therefore $p_1 = p\,(1 - kgt)$, and $\frac{p_1}{p} = 1 - kgt = c$, a constant if variations in g are neglected.

Therefore taking a layer A of any thickness x, and

dividing it into m layers of very small thickness t, we find that if p_n is the pressure above the nth layer,

$$\frac{p_n}{p_{n-1}} = \frac{p_{n-1}}{p_{n-2}} = \ldots = c.$$

And if P and p are the pressures above and below the layer A, $p = Pc^m$; and if d, D be the corresponding densities, $d = Dc^m$.

Therefore in an atmosphere at uniform temperature and at rest, the density and pressure diminish in geometric progression when the height increases in arithmetic progression.

§ **11. Boyle's Law.** *When the temperature of a given mass of gas remains constant, the volume varies inversely as the pressure.*

This law may be verified for air by the following experiments.

I. To verify Boyle's Law for pressures less than that of the atmosphere.

A straight glass tube, closed at one end, is gauged by weighing the quantities of mercury which are required to fill it up to marks, etched on the side of the tube. It is then filled with mercury and inverted in a deep cistern, and a small quantity of air is passed up the tube. The mercury column in the tube falls, and can be adjusted (by raising or depressing the tube in the cistern) till its top coincides with one of the marks.

Let h_1 be the height of the mercury column in the tube, H_1 the height of a barometer column, obtained by reading an adjacent barometer.

Then if p_1 be the pressure of the air in the tube,

$$p_1 + \rho g h_1 = \rho g H_1, \text{ or } p_1 = \rho g (H_1 - h_1).$$

Also V_1 the volume of the air is known.

Now let the tube be depressed in the cistern till the air occupies a volume V_2, and let h_2 be the height of the column in the tube, H_2 the height of the barometer, p_2 the pressure of the air.

Then $$p_2 = \rho g(H_2 - h_2).$$

It is found that if the temperature of the gas has remained the same in both operations, $p_1 V_1 = p_2 V_2$.

Therefore the product of the pressure and volume of a given mass of air remains the same for all pressures less than that of the atmosphere if the temperature is constant.

II. *To verify Boyle's Law for pressures above atmospheric pressure.*

Mercury is poured into a U-tube the longer arm of which is open, and by inclining the tube the air in the closed arm is brought into free communication with the atmosphere, thus taking the same pressure.

More mercury is now poured into the long arm, compressing the confined air, and causing the columns of mercury in the two tubes to assume differents heights. If h is the height of the free surface in the open tube above the other surface and π the pressure of the atmosphere, $\pi + \rho g h$ is the pressure of the confined air; the volume must be obtained by gauging the tube. The volumes occupied by the air under different pressures having been observed, it is found that Boyle's Law is satisfied to a high degree of accuracy.

For air, hydrogen, nitrogen, and other gases which are not readily liquefied the discrepancies from this law at moderate pressures are so slight that they can only be detected by most careful experiments. For gases which are more easily liquefied, as carbonic acid, the law at ordinary temperatures is a rough approximation to the truth.

Manometers.

Manometers, or instruments for direct measurement of pressure, are of two kinds, open and closed.

The open manometer consists of an open U-tube containing mercury, one arm of the tube being connected to the enclosure within which the pressure is to be measured, and the other arm opening into the air.

If the free surface of mercury in the open arm is at a height h above that in the closed arm, the pressure is $\pi + \rho g h$, where π is the pressure of the atmosphere.

The closed manometer differs from the above in having one arm of the tube closed at the top, instead of opening into the air.

The tube is gauged and the volume occupied by the contained air at standard pressure is noted. Thence the pressure of the air when occupying a different volume can be deduced from Boyle's Law, and if p be this pressure, h the same difference of heights as before, $p + \rho g h$ is the pressure to be measured.

Law of Charles. If the temperature of a mass of gas which obeys Boyle's Law varies, it is found that

$$pv = R(1 + \alpha t),$$

where p is the pressure, v the volume, and t the temperature.

The coefficient α is approximately the same for all such gases, and is about $\frac{1}{273}$ when temperature is measured on the centigrade scale.

Absolute temperature T on the air thermometer is defined

by the relation $T = t + 273$, t being the temperature on the centigrade scale.

Thus the zero of absolute temperature is -273° C, and the laws of Boyle and Charles are expressed by the relation

$$pv = RT,$$

where R is a constant which depends on the nature of the gas.

Law of the mixture of gases.

Two portions of gas at the same temperature, but at different pressures, are mixed. To show that if p_1, p_2 are the initial pressures of the gases, v_1 and v_2 the initial volumes, V the volume of the mixture under a pressure P,

$$PV = p_1v_1 + p_2v_2.$$

If the pressure of the second gas were p_1, its volume would be v', where $p_1v' = p_2v_2$. Now, when two gases at the same pressure p are mixed, the pressure of the mixture is found to be p, if the total volume of the gases is unchanged. Hence, under a pressure p_1, the volume of the mixture is $v_1 + v'$; and since the mixture obeys Boyle's Law, the volume V under a pressure P is given by the relation

$$PV = p_1(v_1 + v') = p_1v_1 + p_2v_2.$$

§ 12. Pumps.

The suction pump (Fig. 145) consists of a vertical cylindrical barrel opening below into a smaller cylindrical tube called the suction tube.

A piston rod C worked by the pump handle moves up and down in the barrel, and the suction tube dips below the surface of the water that is to be pumped up. At the junction of the pump barrel with the suction tube there is a valve D opening upwards, and an aperture in the piston is covered by another valve *E*, which also opens upwards.

The following is the mode of action of the pump. Initially the valves are closed, the air in the barrel and suction tube is at atmospheric pressure, and the water surface is at the same level inside and outside the suction tube.

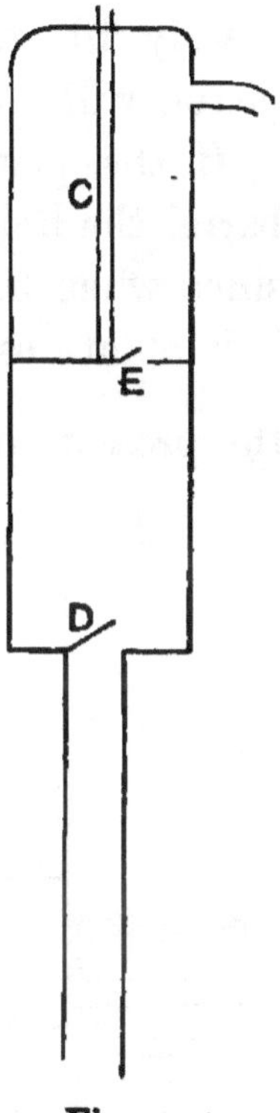

Pumping begins by raising the piston from the bottom of the barrel. This increases the volume of air in the barrel and lowers its pressure; consequently the lower valve opens and air passes from the suction tube into the barrel, and the diminution of pressure in the suction tube causes the water to rise into it, until the upward stroke has been completed.

When the piston descends the pressure below it increases, the upper valve opens, the lower valve shuts, air is exhausted from the barrel, and the level of the water in the suction tube remains stationary.

Fig. 145.

In the second upward stroke of the piston, the water rises to a greater height, and after a certain number of strokes reaches the pump barrel. The length of the suction tube must not exceed the height of the water barometer (about 34 feet), since the pressure at the foot of the column must be that of the atmosphere. When the water has risen into the barrel, the action changes; as the piston descends, water as well as air escapes through the upper valve, and at the end of the stroke all the water in the pump barrel is above the piston. When the piston begins to rise the valve closes, and the water above the piston is carried up and discharged by the mouth of the pump at the end of the stroke.

If the course of the piston carries it more than 34 feet above the water in the well the pump barrel will not completely fill with water during the upward stroke, and the piston will not work to full advantage.

If the piston does not descend to the bottom of the barrel the height to which the water can rise is diminished, since when the piston is in its lowest position the air below it is at atmospheric pressure π, and when the piston rises the pressure of this air cannot fall below $\frac{\pi h}{H}$, H being the height of the barrel, h the height untraversed by the piston. The maximum possible height to which such a pump can raise water by suction is

$$34 - \frac{\pi h}{H} \text{ feet.}$$

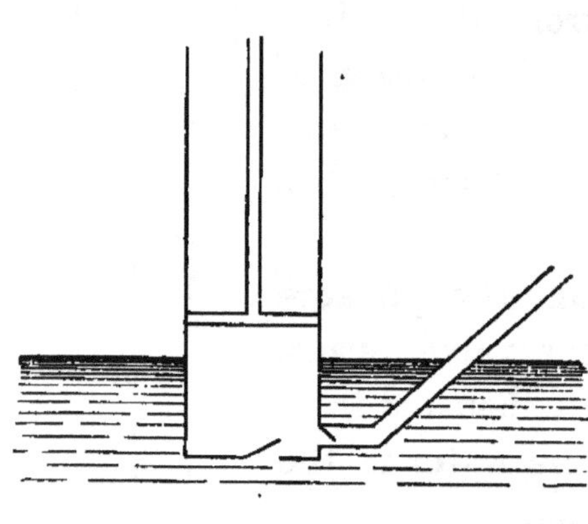

Fig. 146.

The Force Pump (Fig. 146). This consists of a pump barrel, dipping into water, with a valve at the bottom opening inwards and a valve at the side opening outwards. A piston works in the pump barrel and, when it descends, closes the bottom valve and forces water out of the barrel through the side valve. When the piston rises, the side valve closes and the lower valve opens, admitting water from the reservoir into the barrel.

Air Pump.

The principle of the Air Pump is precisely that of the

Suction Pump; the receiver A, from which air is pumped, corresponds to the well; the tube B corresponds to the suction tube, and the arrangement of the barrel and piston is the same as in the suction pump. When the piston descends, the lower valve closes and the upper one opens, air escaping through it from the barrel. When the piston ascends the upper valve closes, the lower one opens, and air passes from the receiver into the barrel.

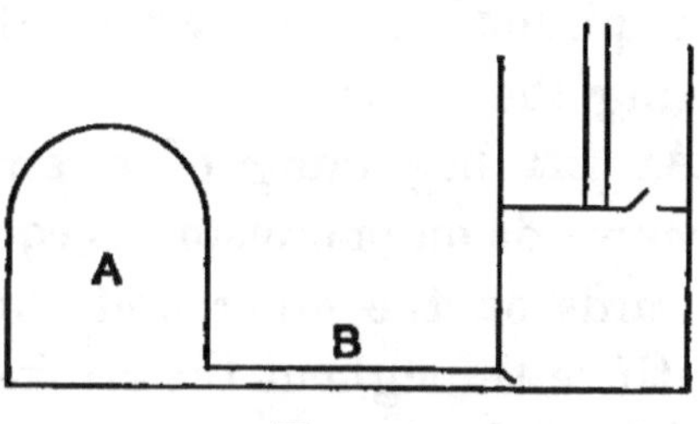

Fig. 147.

We can calculate the theoretical rate of exhaustion.

Let V be the volume of the barrel, V' that of the receiver and the tube leading to it, P the pressure in the receiver when the piston is in its lowest position.

When the piston has been raised the air which originally occupied a volume V occupies a volume $V+V'$, and by Boyle's law its pressure is $\frac{PV}{V+V'}$.

The descent of the piston does not alter the pressure in the receiver. Therefore each complete stroke diminishes the pressure in the receiver in the ratio $\frac{V}{V+V'}$, and after n strokes the pressure in the receiver is $P\left(\frac{V}{V+V'}\right)^n$.

In the pump here described, if P be the pressure of the atmosphere, S the area of the piston, p the pressure in the receiver, the total force required to raise the piston is

$(P-p)S$, and when exhaustion has proceeded pretty far, this force is inconveniently great.

This difficulty may be overcome by having two barrels to the pump, the pistons in which are worked by a rack and pinion motion, which depresses one piston while raising the other.

At the beginning of a stroke the resultant pressure downwards on one piston is equal to the resultant pressure upwards on the other, and though this equality does not continue throughout the stroke, the arrangement eases the action of the pump.

This pump also produces more rapid exhaustion, for air is always passing from the receiver into one barrel or the other.

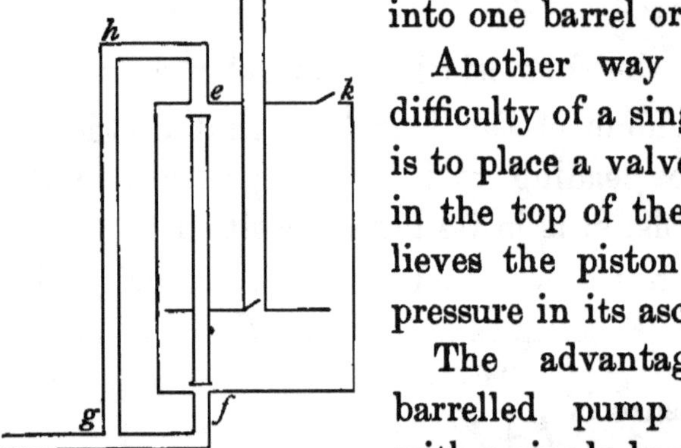

Fig. 148.

Another way of obviating the difficulty of a single barrelled pump is to place a valve opening upwards in the top of the barrel. This relieves the piston from atmospheric pressure in its ascent.

The advantages of a double barrelled pump can be obtained with a single barrel as follows:—

The piston rod is hollow, and the piston valve opens into it; a valve at the top of the barrel opens upwards.

Two openings in the barrel at *e* and *f* communicate through the tubes *ghe*, *gf* with the receiver, and are closed by stoppers at the ends of a metal rod which passes through the piston and is carried by friction with it in its vertical movement. When the piston ascends, the opening *e* and the valve in the piston rod are closed, the opening *f* and the valve *k* are open. Thus air flows in from the receiver below the piston and passes out of the barrel above into the room. When the piston descends, air from the receiver enters above the piston, and below the piston is expelled into the room.

Compressing Pump.

The pump can also be used for the compression of air; it is only necessary that the valves should be reversed, so that in the ascent of the piston atmospheric air is admitted into the barrel, and in the descent the air is pushed into the receiver.

If V is the volume of the receiver of a single-barrelled pump, V' that of the barrel, and P the atmospheric pressure, the pressure in the receiver after one stroke is

$$\frac{P(V+V')}{V}.$$

After n strokes, a mass of air occupying a volume nV' at atmospheric pressure has been passed into the receiver, in addition to the volume V which it originally contained.

Therefore the pressure after n strokes is $\dfrac{(V+nV')P}{V}$.

Examples.

1. The apparent weight of a piece of platinum when immersed in water is 20·6 gr. When immersed in mercury it is 8 gr. The density of mercury being 13.6, find the volume and density of the platinum.

If V is the volume and ρ the density, $V(\rho-1)g$ is the apparent weight in water and $V(\rho-13{\cdot}6)g$ is the apparent weight in mercury.

Therefore $V(\rho-1)=20.6$, and $V(\rho-13{\cdot}6)=8$.

Therefore $V=1$ and $\rho=21{\cdot}6$.

2. 330 ccm. of cork float in a liquid of density 1.2 with 77 ccm.

immersed. Find the density of the cork and the added weight that would totally submerge it.

The mass of the cork is equal to the mass of liquid displaced, that is to 77×1.2 gr. $= 92.4$ gr.

And the density of the cork is $\frac{92.4}{330} = \frac{8.4}{30} = \frac{7}{25}$.

3. The barometer stands at 30 inches. Find in tons' weight the pressure per linear horizontal foot on the sides of a rectangular tank 10 feet deep filled with water, the specific gravity of mercury (i. e. its density relative to water) being 13.6, and the mass of a cubic foot of water being 1000 oz.

The pressure of the atmosphere is equal to that of a column of water $\frac{30 \times 13.6}{12} = 34$ feet high.

The pressure at the centroid of a vertical face is that due to $34 + 5$ or 39 feet of water.

And by § 8 the total pressure per lineal horizontal foot is

$$39 \times 10 \times 1000 \text{ oz. weight}$$
$$= 10\tfrac{395}{448} \text{ tons' weight.}$$

4. A tube closed at the upper end and containing air dips into a deep cistern of mercury, the mercury standing at the same level inside and outside. If the height of the portion of the tube above the mercury is equal to the height of the barometer (30 inches), find how much the tube must be raised that the contained air may occupy 40 inches of the tube.

The original volume of the air is $\frac{3}{4}$ of the final volume.

Hence the final pressure is $\frac{3}{4} \times 30$ inches $= 22\frac{1}{2}$ inches.

Therefore the tube must be raised through $40 - 22\frac{1}{2} = 17\frac{1}{2}$ inches.

5. A barometer which has a little air in it reads 29.6 inches, the end of the tube being 6 inches above the top of the mercury, when a standard barometer reads 30. What is its reading when the standard registers 29?

The pressure of the air above the mercury is $30 - 29.6$ when its volume is 6 (arbitrary units).

Let x be the reading required.

Then $29 - x$ is the pressure of the air above the mercury, and $35.6 - x$ is its volume.

Therefore by Boyle's Law

$$(35{\cdot}6 - x)(29 - x) = 6 \times .4,$$

$$\text{or} \quad x^2 - 64.6x + 1030 = 0.$$

$$x = 32{\cdot}3 \pm 3{\cdot}65 \text{ approximately.}$$

The upper sign gives the air a negative volume and pressure, which is impossible.

Therefore $x \doteq 28.65.$

6. Taking a cubic foot of water as 1000 oz., find the total pressure of the water arising from its weight on a side of a cistern 7 feet wide and 8 feet deep, if the cistern is filled with water.

What horizontal line would divide the side into two parts so that the total pressure on each would be the same?

7. A body consists of an alloy of two metals of specific gravities s_1, s_2 respectively; its weight in vacuo is w and in water is w'. Show that the proportion of the two metals (by volume) is

$$s_2 w' - (s_2 - 1) w : (s_1 - 1) w - s_1 w'.$$

8. A sphere whose internal radius is 1 foot contains mercury which covers $\frac{3}{4}$ of the vertical diameter; find the resultant pressure on a horizontal plane through the centre of the sphere, assuming that a cubic inch of mercury weighs 3400 grains.

9. If in 10 strokes the mercurial gauge of an air-pump fell from 30 inches to 15 inches, what would have been its fall after the first 5 strokes?

10. A square whose side is 8 ft. long has its plane vertical and its upper edge on the surface of the water in which it is immersed; find the resultant pressure on one face of it. If the square is fixed and the surface of the water raised a foot, what would now be the magnitude of the resultant pressure?

Find also the centre of pressure in each case.

11. A spherical shell made of material of density ρ floats half immersed in water. Find the ratio of the internal to the external radius.

12. A cubical vessel is filled with two liquids of density 1 and 0.8. They do not mix and their volumes are equal. Find the ratio of the resultant pressure on the upper to that on the lower half of one of the vertical faces of the cube.

13. 1000 cubic inches of air under a pressure of 20 lbs. per square inch are mixed with 800 cubic inches of air under a pressure of 15 lbs. per square inch.

Find the pressure of the mixture when it has a volume of 1500 cubic inches, the temperature being the same in all cases.

14. A hydrometer floating in a liquid of density 1.1 has 5 cm. of the stem above the surface; when it floats in liquid of density 1.2 it has 6 cm. of the stem above the surface. How much of the stem will be above the surface in a liquid of density 1.3?

15. 3 litres of liquid of density 0.8 are mixed with 5 litres of another liquid of density 1.04. Find the density of the mixture, assuming (1) that there is no contraction, (2) that there is a contraction of 5 p. c. of the joint volumes.

16. A wooden vessel 6 in. square and 6 in. in height with a neck 2 in. square and 6 in. in height is filled with water. Find the centre of mass of the water, and the total pressure on the base of the vessel.

17. A piece of silver and a piece of copper, fastened to the ends of a string passing over a pulley, hang in equilibrium when suspended in a liquid of density 1.15. Determine the relative volumes of the masses, the densities of silver and copper being 10.47 and 8.89 respectively.

18. A piece of cork floats in a beaker of liquid. If the beaker is placed under the receiver of an air-pump and the air above it exhausted, will the cork rise or sink in the liquid?

19. A vessel shaped like a portion of a cone is filled with water. It is 1 inch in diameter at the top, and 8 inches at the bottom, and is 12 inches high. Find the pressure (in pounds per square inch) at the centre of the base and the total pressure on the base.

(A cubic foot of water weighs 1000 oz.)

20. A forcing pump the diameter of whose piston is 6 inches is employed to raise water from a well to a tank. If the bottom of the piston is 20 feet above the surface of the water in the well and 100 feet below the surface in the tank, find the least force which will (1) raise, (2) depress the piston; friction and the weights of the valves being neglected.

21. A Nicholson's Hydrometer weighs 8 oz. The addition of 2 oz. to the upper pan causes it to be sunk in one liquid to the mark, while 5 oz. are required to produce a like effect in another liquid. Compare the densities of the liquids.

If the density of the weights employed is 8, what weight must be placed in the lower pan in water to produce the same effect as 2 oz. in the upper pan?

22. A Nicholson's Hydrometer when loaded with 200 grains in the upper pan sinks to the marked point in water; a stone is placed in the upper pan, and the weight required to sink it to the same point is 80; the stone is then placed in the lower pan and the weight required is 128 grains. Find the density of the stone.

23. A piece of lead and a piece of sulphur are suspended by fine strings from the extremities of a balance beam and just balance each other in water. Compare their volumes, their densities being respectively 11·4 and 2. Which will appear the lighter in air and what weight must be added to it to restore equilibrium?

24. A siphon is filled with water and inverted into a vessel of liquid of density 1·6. What is the condition that the liquid may flow through the siphon?

25. If in Bramah's Press a (total) pressure of 1 ton is produced by a force of 5 lbs., and the diameter of the pistons are as 8 : 1, find the ratio of the arms of the lever employed to work the piston.

26. A piece of silver and a piece of gold are suspended from the ends of a balance beam which is in equilibrium when the silver is immersed in alcohol and the gold in nitric acid. The densities of gold, silver, nitric acid, and alcohol being 19·3, 10·5, 1·5 and 0·85 respectively, compare the masses of gold and silver.

27. A piece of wood floats partly immersed in water, and oil is poured on the water till the wood is completely covered. Explain whether this makes any change (and if so whether there is an increase or decrease) in the quantity of wood immersed in water.

28. If the height of the barometer changes from 29·55 to 30·33 inches, what is the change in the weight of 1000 cubic

inches of air, assuming that 100 cubic inches of air weigh 31 grains at the former pressure, the temperature being constantly at 0° C ?

29. 10 ccm. of air are measured at atmospheric pressure, when introduced into a barometer vacuum they depress the mercury which previously stood at 76 cm., and occupy a volume of 15 ccm. By how much has the mercurial column been depressed ?

30. A faulty barometer tube 33 inches long contains air at the top and consequently reads 28 inches when the true pressure of the atmosphere is 29. Find its reading when the pressure of the atmosphere is 28 inches.

31. In an air-pump with one barrel the volume of the receiver is 10 times that of the barrel, and the lower $\frac{1}{10}$ of the barrel is not cleared by the piston. Find (1) the pressure after two complete strokes, (2) the lowest pressure that can be obtained with the pump.

32. A piece of lead weighing 17 grams and a piece of sulphur have equal apparent weights when suspended from the pans of a balance and immersed in water. When the water is replaced by alcohol of density 0.9, 1.4 grams must be added to the pan from which the lead is suspended to restore equilibrium. Find the weight and volume of the sulphur, the density of lead being $11\frac{1}{3}$.

33. Two dock gates close a channel 12 feet wide, the depth of water on one side of the gate is 3 feet and on the other 15 feet. Find in tons' weight the force that must act on either gate to prevent them from opening.

CHAPTER IX.

CAPILLARITY.

§ 1. WHEN two particles of liquid are separated by less than a very small distance a (called the range of molecular action) they exert a certain force on each other. Thus if m be a particle of liquid round which as centre a sphere P is described with radius a, the particles which exert force on m lie within the sphere P.

The potential energy of the liquid is equal to the work that its forces would do in separating its particles to such an extent that the distance between any two of them is greater than a.

Let the external forces acting on the particles of the liquid be insignificant in comparison with the molecular or internal forces; there may however be a uniform pressure of any magnitude on the surface of the liquid. The internal forces are supposed to form a conservative system.

In considering the work done in removing m from the action of neighbouring particles two cases arise, according as the sphere of action of P is entirely within the surface, or meets it.

In the first case the potential energy of each particle is a constant U; in the second case the energy differs from U by a quantity which depends on the distance of m from the surface.

Thus the potential energy of all particles at a given distance from the surface is the same, and if the distance is very small their number is proportional to the area of the surface.

Therefore the potential energy of a liquid containing n particles is proportional to $nU + kS$, where S is the area of the surface, and k is a constant which depends on the nature of the liquid but not on the form of the surface.

A system of particles tends to have as little potential energy as possible, and since kS is the only term in the energy which can vary, it will become as small as possible.

Therefore a liquid when free from external forces other than a uniform pressure over its surface, assumes such a form that its surface is either of maximum or of minimum area, according as k is negative or positive.

Thus a small drop of mercury assumes a spherical form, for the surface of a sphere is less than the surface of any other solid of the same volume. Pure water in contact with glass tends to assume a maximum surface, spreading out into a thin film.

The energy which a liquid possesses in virtue of its surface is called its Superficial Energy. Different liquids possess, for equal surfaces, different amounts of superficial energy, and the superficial energy per unit area is called the Constant of Capillarity.

Since the surface tends to be as small as possible, it offers resistance to extension, just as a stretched membrane does.

Phenomena of superficial energy are called capillary (from *capilla*), because they are most conspicuous in very fine tubes.

Let $ABCD$ be a rectangular framework, having the bar BC moveable along AB and CD, and containing a liquid film whose superficial energy is T per unit area.

The loss of potential energy in moving BC through a distance h towards AD is $2hBC.T$, $2hBC$ being the decrease of surface.

A tension $2T.BC$ applied to BC would maintain equilibrium, since the work done against it in this displacement is equal to that supplied by the film. This would be a tension T per unit length of BC on each face of the film.

We may therefore consider the surface of a liquid as a membrane stretched by a tension T; that is, across any line of length l on the surface a force Tl is exerted, uniformly distributed along the line. Hence T is often called the Surface Tension of the liquid.

Examples of capillary phenomena. If a wire ring is dipped into a soap solution and withdrawn, it carries with it a film of liquid whose surfaces are plane. If a loop of thread is laid on the film, and the film broken within the loop, the remainder of the film contracts until the loop forms a circle. As the circle has a larger area than any other curve of the same perimeter, the area of the film is the smallest possible.

A needle laid gently on water will sometimes float on it, being supported by the tension of the surface.

If a mixture of alcohol and water is made to have the density of oil, and by means of a pipette a drop of oil is placed inside the mixture, the drop assumes a spherical form. Here the fluid pressure balances the force of gravity, and the shape of the drop is determined by surface tension only.

§ 2. Pressure within a spherical drop of liquid.

If a thin membrane with a fixed boundary is stretched, and unequal pressures are applied to its faces, the membrane becomes curved, and the curvature increases with the difference of the pressures.

Hence the pressure within a drop of liquid exceeds the external pressure by an amount which depends on the curvature of the drop.

Let p be the pressure within a spherical drop of radius r, T the surface tension, P the atmospheric pressure.

Consider the conditions of equilibrium of half the drop, bounded by a plane through the centre O perpendicular to a radius On.

The hemisphere is in equilibrium under the external atmospheric pressure, the pressure across its plane circular boundary, the tensions at the perimeter of this boundary, and its weight.

The resultant of the tensions is $2\pi rT$ parallel to nO.

The resultant of the pressures on the plane boundary is $\pi r^2 p$, parallel to On, and that of the pressures on the curved boundary is $\pi r^2 P$, parallel to nO (Chap. VIII. § 7).

The weight is $\frac{2}{3}\pi r^3 \rho g$ and its component along nO can be neglected when r is small.

Therefore the condition of equilibrium is

$$\pi r^2 p = 2\pi r T + \pi r^2 P,$$

$$\text{or} \quad p = P + \frac{2T}{r}.$$

Pressure within a soap bubble.

The bubble is a liquid film each of whose surfaces has a tension T, and as the bubble is very thin we may take the radii of the faces as equal, and neglect the weight of the bubble.

Hence the pressure within the film is $P + \frac{2T}{r}$, and the pressure within the bubble is $P + \frac{4T}{r}$.

§ 3. **Conditions of contact of three fluids,** *e.g.* **water, mercury, and air.**

If a drop of water rests on mercury in air, we have three surfaces, mercury-water, water-air, and air-mercury, intersecting in a line of which a very small part L, of length h, can be considered as straight.

Let the line be perpendicular to the plane of the paper, and let OA, OB, OC be tangents in the plane of the paper to the surfaces. The surfaces can be regarded as stretched membranes in which the tensions are T_A, T_B, T_C respectively.

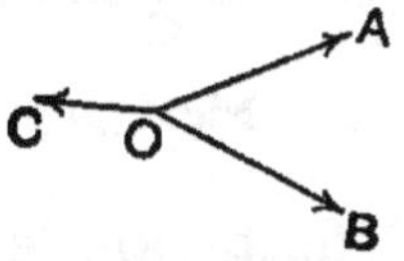

Fig. 149.

A cylinder of liquid of height h, which has L for its axis, is in equilibrium under the pressure of the surrounding fluid and the forces hT_A, hT_B, hT_C. But by diminishing the base of the cylinder we can make the resultant pressure on its surface diminish indefinitely, and therefore the forces hT_A, hT_B, hT_C, acting at O, must satisfy the condition of equilibrium. Hence $T_A : T_B : T_C :: \sin BOC : \sin COA : \sin AOB$, and the angles AOB, BOC, COA can be found when T_A, T_B, and T_C are known.

If one of the tensions is greater than the sum of the other two, the fluids cannot rest in contact along a line; the spreading of oil on water is a consequence of this.

The above argument also holds when two fluids are in contact with a solid.

A solid in contact with a liquid in air is generally wetted, the liquid-air and solid-air surfaces being in contact; if OC is in the solid-air surface, OA and OB coincide and are in the same straight line with OC; hence $T_C = T_A + T_B$.

In some cases, as when mercury is in contact with glass, the liquid does not wet the solid, but the solid-air and liquid-air surfaces intersect at an angle.

Then if OA is in the liquid-air surface, the conditions of equilibrium give the relation

$$T_C - T_B = T_A \sin AOB.$$

Hence AOB, or the angle of contact of the solid and liquid is known when T_A, T_B, T_C are known. Hence under conditions for which T_A, T_B, T_C are constant, the angle AOB is a fixed angle.

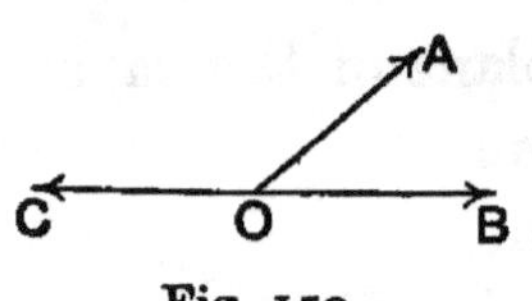

Fig. 150.

It follows that when a liquid is in contact with a solid its surface makes a definite angle with that of the solid, and the potential energy of the liquid is as small as it can be consistently with this condition.

Equilibrium of a heavy drop.

In this case the drop still tends to possess a minimum surface, but its centroid also tends to lie as low as possible, and by the Principle of Virtual Work the form of the drop is such that the work done by gravity in a very small change of form is equal and opposite to that done by the surface-tension. Thus, while very small drops of mercury are approximately spherical, a large drop, lying on a horizontal plane, assumes a flattened form.

§ 4. Elevation of Liquids in capillary tubes.

When a tube is immersed vertically in water, the free surface rises higher in the tube than outside it, and the narrower the tube, the greater is the elevation of the water.

Let r be the radius of the tube, h the difference of the levels of the free surfaces within and without the tube, ρ and T the density and surface tension of the liquid.

When the tube is very narrow we may take the curved surface, or meniscus, to be a sphere.

If P is the atmospheric pressure, $P - \frac{2T}{r}$ is the pressure just within the meniscus, and $P - \frac{2T}{r} + \rho g h$ is the pressure at the level of the free surface outside the tube; but this is equal to P.

Therefore $$\frac{2T}{r} = \rho g h,$$

or $$T = \tfrac{1}{2}\rho g r h.$$

Hence the elevation of liquid in a tube is proportional to the surface tension, and inversely proportional to the radius of the tube and to the density of the liquid.

This affords the simplest method of measuring the surface tension of such a liquid as water.

The radius r of the tube may be determined by measuring the length l of a column of mercury in the tube, and determining the mass m of the mercury. Taking the density of mercury ρ as about 13·55 at ordinary temperatures we have $\pi r^2 l \rho = m$, from which r can be found.

Depression of mercury in a tube.

Mercury in contact with glass and air assumes as small a surface as possible, and the pressure within its surface exceeds that of the air. Hence the surface of mercury is depressed in the tube and has its convexity upwards.

If a is the angle of contact, $\frac{r}{\cos a}$ is the radius of the surface, and the pressure within the meniscus is

$$P + \frac{2T}{r}\cos a.$$

Hence the meniscus is at a depth h below the free surface in the cistern, where

$$2T\cos a = \rho g r h.$$

Thus the pressure of the atmosphere determined by a mercury barometer is less than the true pressure, owing to the surface tension of the mercury.

§ 5. Elevation of liquid between two parallel plates.

It is first necessary to find the tension T of a cylindrical membrane of radius r, filled with liquid at pressure p and subjected to an external pressure P.

Bisect the cylinder by a plane L through its axis and consider the portion intercepted between this plane, the curved surface, and two planes perpendicular to the axis at a distance l apart.

Resolve the acting forces along a perpendicular to L.

The resultant of the tensions in the membrane is $2\,Tl$.

The resultant of the internal pressures is $2\,rlp$, and of the external pressures $2\,rl\,P$ (Chap. VIII. § 7).

Therefore $$p = P + \frac{T}{r}\,.$$

When two parallel plates are immersed vertically, at a small distance $2\,r$ apart, in a liquid which wets them, the liquid surface, except near the edges of the plates, is approximately a horizontal cylinder of radius r.

Hence if the liquid wets the plates the pressure within the meniscus is $P - \frac{T}{r}$, and at a depth h below the meniscus it is $P - \frac{T}{r} - \rho g h$.

This is equal to P if $T = \rho r g h$.

Hence the elevation of the liquid in this case is half of the elevation in a tube of radius r.

Apparent attraction of the plates.

The external pressure on the free portion of either plate

is P, and the internal pressure at a height x above the surface in the cistern is $P-\rho g x$.

Hence at a height x there is an excess of external pressure $\rho g x$.

If the plates are rectangular with one pair of edges horizontal, and if h is the height to which the water rises, b the width of the plate, the area immersed is bh, and the resultant pressure is $\frac{1}{2} bh^2 \rho g$.

Thus the pressures urge the plates towards one another, and the plates apparently attract one another.

The resultant pressure may also be written

$$\frac{1}{2}\frac{bT^2}{\rho g r^2}.$$

The attraction between the plates thus increases rapidly as the distance between them diminishes; two wet plates fairly pressed together require considerable force to separate them. When the liquid is depressed between the plates, there is also an apparent attraction.

Illustrations.

1. A thin flexible string forming a loop of length $2\pi r$ is laid on a plane film, and the film within the loop is broken; find the tension X of the string.

Let T be the surface tension. The string becomes a circle of radius r, each portion of it being pulled outwards by a tension $2T$ (for the film has two surfaces) per unit length. A part of the string which forms a semi-circle is in equilibrium under these forces and the tensions X at its extremities, and by Chap. VIII. § 7 the liquid tensions are equivalent to a single force $4rT$ perpendicular to the diameter which bounds the semi-circle.

Therefore $\qquad 2X = 4rT, \quad$ or $\quad X = 2rT.$

2. A liquid film OAB of surface tension T is bounded by two wires OA and OB, each of length l, and a light inextensible string AB of length $\frac{\pi l}{3}$. The wire OA is fixed horizontally and OB can

turn freely round O. Prove that if OB rests at an angle 60° to the horizon its weight is $8Tl$.

Complete the parallelogram $OADB$.

BA subtends an angle $60°\left(=\frac{\pi}{3}\right)$ at O and at D. Also $BA = \frac{\pi l}{3}$ and $BD = DA = l$. Therefore D is the centre of the circle of which AB is an arc.

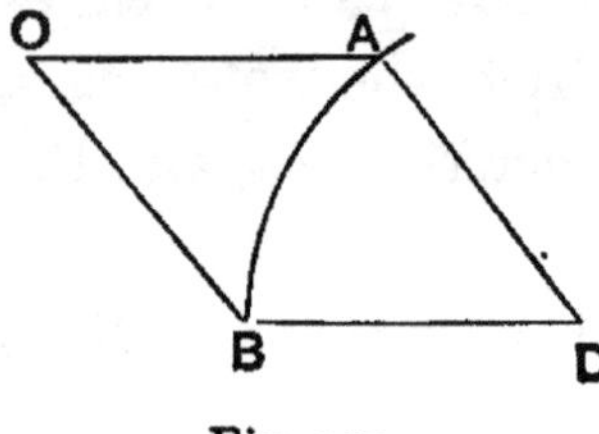

Fig. 151.

The forces which act on OB are its own weight W, the tension $2Tl$ perpendicular to OB, the resistance at O, and the tension t in the string at B. By the last example, $t = 2Tl$.

Hence taking moments round O, $W = 8Tl$.

3. Two soap bubbles of radius r_1, r_2 rest in contact along a circle. Show that the film between the bubbles is a sphere of radius $\frac{r_1 r_2}{r_1 \sim r_2}$.

Let P be the atmospheric pressure.

$$P + \frac{4T}{r_1} \text{ is the pressure within the first bubble;}$$

$$P + \frac{4T}{r_2} \text{ is the pressure within the second.}$$

The difference of pressures is $4T\left(\frac{1}{r_1} \sim \frac{1}{r_2}\right)$.

Therefore the radius r of the bounding surface is given by

$$\frac{1}{r} = \frac{1}{r_1} \sim \frac{1}{r_2}.$$

At a point common to the three spheres, the tensions are equal and in equilibrium. Hence the spheres cut at an angle of 120°.

If $r_1 = r_2$, the bounding surface becomes a plane.

Examples.

1. In a tube whose internal diameter is 1 mm. a certain solution which wets the side of the tube rises to a height of 10 mm. Calculate the pressure within a bubble of this solution of 1 cm.

radius, the atmospheric pressure being 76 cm. of mercury, the density of the mercury 13·6, and the density of the solution 1·1.

2. A soap bubble of radius r rests within a bubble of radius $2r$. If T is the surface tension, d the density of air under atmospheric pressure, find the total mass of air inside the larger bubble.

3. Two vertical tubes of radii r_1, r_2, are connected below by a pipe and contain liquid of density ρ and surface tension T. Prove that the difference of the heights of the liquid columns is

$$\frac{2T}{\rho g}\left(\frac{1}{r_1} \sim \frac{1}{r_2}\right).$$

4. A hydrometer consisting of a glass cylinder of radius r, loaded below, floats in a liquid of density ρ and surface-tension T.

Prove that if the liquid were free from surface-tension, the hydrometer would rise through a height $\frac{2T}{r\rho g}$.

5. A vertical framework is formed by two horizontal wires of length l, and two light flexible strings of length l', attached to the end of the wires. The upper wire is fixed and the framework encloses a soap film. Calculate the weight of the hanging wire if the strings at their lower extremities make an angle 30° with the vertical.

CHAPTER X.

Units and their Dimensions.

§ 1. Units.

Every measurable quantity is expressed by two factors, one a pure number or numeric, the other the unit in terms of which the quantity is measured. Thus in the expression 10 lb., 10 is the number and the lb. is the unit.

It has been already seen that physical quantities are most easily dealt with when they are measured as multiples of units which are derived from the units of mass, length, and time. A system of units connected in this way is called an **Absolute System of Units**, and any one of the units is called an absolute unit.

The units of mass, length, and time are perfectly arbitrary, and if they are altered the other units alter too. The question that we shall now consider is—In what way do the derived units alter, when the fundamental units are changed?

If (l) is the unit of length, e.g. the foot or centimetre, $l(l)$ is the full expression of a length containing l units.

Similarly, if (t) is the unit of time and (m) the unit of mass, a time and a mass hitherto denoted by t and m are fully described by the symbols $t(t)$ and $m(m)$.

If a new unit of length (L) is chosen, and the length $l(l)$ contains L of the new units,

$$L(L) = l(l). \qquad (1)$$

Let the units of length, time, and mass be changed from (l) to $x(l)$, from t to $y(t)$, and from (m) to $z(m)$, x, y, z being any numbers or fractions. We shall determine the changes in the derived units.

Units of Area and Volume.

If (s) is the unit of area when (l) is the unit of length, the rectangle contained by the lines $h(l)$, $k(l)$ is $hk(s)$. When the unit of length is changed to $x(l)$ or (L), the lines become $\frac{h}{x}(L)$, $\frac{k}{x}(L)$. And the rectangle becomes $\frac{hk}{x^2}(S)$, where (S) is the new unit of area.

Therefore $\frac{hk}{x^2}(S) = hk(s)$, and $(S) = x^2(s)$.

Hence the unit of area is proportional to the square of the unit length. Similarly the unit of volume is proportional to the cube of the unit length.

Unit of Velocity.

Let (v) be the unit of velocity in the system (l), (m), (t), i.e. the system derived from (l), (m), (t) as fundamental units.

If a body traverses a distance $l(l)$ in a time $t(t)$ its velocity is $v(v)$, where $v = \frac{l}{t}$.

Now let the units be changed to (L) and (T), and let (V) be the new unit of velocity.

Then $l(l) = \frac{l}{x}(L)$ and $t(t) = \frac{t}{y}(T)$, and the new measure of the velocity is V, where $V = \frac{l}{x}\frac{y}{t} = \frac{y}{x}v$.

Also $V(V) = v(v)$. Therefore $(V) = \frac{x}{y}(v)$.

Thus the unit of velocity is directly proportional to the

unit of length and inversely proportional to the unit of time.

Unit of Acceleration.

Let $v(v)$ be the velocity acquired in time $t(t)$ by a body. Its acceleration is $a(a)$, where (a) is the unit of acceleration, and $a = \frac{v}{t}$.

In the new system the acceleration is $A(A)$, and

$$A = \frac{V}{T} = \frac{y^2}{x}\frac{v}{t} = \frac{y^2}{x}a.$$

Therefore since $A(A) = a(a)$, $(A) = \frac{x}{y^2}(a)$.

Unit of Force.

Let a force $f(f)$ be required to communicate an acceleration $a(a)$ to a mass $m(m)$.

Then $$f = ma.$$

When the units are changed, let the same force be denoted by $F(F)$.

Then $F = MA = \frac{y^2}{zx} \cdot ma = \frac{y^2}{xz}(f)$.

And since $F(F) = f(f)$, $(F) = \frac{xz}{y^2} f$.

Unit of Density.

Let a volume $u(u)$ contain a mass $m(m)$; the density is $d(d)$, where $d = \frac{m}{u}$.

Changing the units, let $U(U)$ be the new measure of the volume, $D(D)$ the density.

Then $D = \frac{M}{U} = \frac{mx^3}{zu} = \frac{x^3}{z}.d$.

And since $D(D) = d(d)$, $(D) = \frac{z}{x^3}(d)$.

The numbers which specify a strain and an angle are the same whatever system of units we adopt.

When the alteration of a fundamental unit (A) in the ratio $1 : p$ alters a derived unit (B) in the ratio $1 : p^m$, (B) is said to be of dimensions m in (A).

Thus the unit of velocity is of dimensions 1 in length and -1 in time.

Let (B) and (C) be two units which are of dimensions m and n respectively in (A), and let them become (b) and (c) when (A) is changed in the ratio $1 : p$.

Then $(b) = p^m (B)$, and $(c) = p^n (C)$; and by multiplication $(b)(c) = p^{m+n} (B)(C)$.

Thus the product of two units has dimensions, which may be obtained by adding the dimensions of the units.

In the remainder of this chapter we shall generally use small letters to denote numerics, and capitals to denote the corresponding unit.

The units already found can be denoted as follows :—

Length	L.
Time	T.
Mass	M.
Velocity	LT^{-1}, or V.
Acceleration	LT^{-2}, or A.
Force	MLT^{-2}, or F.
Density	ML^{-3}, or D.
Angle	Constant.
Strain	Constant.

The dimensions of the products of the units follow from this table. Thus AL and V^2 are both of dimensions L^2T^{-2}.

Each term of any physical equation, e.g. $s = vt + \frac{1}{2}at^2$, is of certain dimensions in the fundamental units. But the equation is true whatever these units may be, and therefore a change of units changes all terms in the equation in the same ratio. This can only be true if the equation is homogeneous in the fundamental units.

Hence all physical equations are homogeneous, and this result may be usefully employed to verify equations, for it is a necessary (but not sufficient) condition of accuracy. It may also be employed to determine the dimensions of physical units.

Other fundamental units might be taken, e.g. V, A, F.

We then have $T = VA^{-1}$, $M = FA^{-1}$, $L = V^2A^{-1}$, whence the relation between any other unit and V, A, F can be found.

Astronomical Units.

If we regard gravitation as a fundamental property of matter, the unit of mass may be made to depend on the units of length and time. The astronomical unit of mass is defined as that which communicates to a mass at unit distance from it the unit of acceleration.

According to this definition and the law of gravitation, a mass m exerts a force $mm'd^{-2}$ or a mass m' at distance d.

Therefore, if a is the acceleration of m', $md^{-2} = a$; since this relation is homogeneous, mass is of dimensions L^3T^{-2} in this system of units.

Substituting L^3T^{-2} for M in the formulae above, we obtain the astronomical dimensions of other units in L and T. The unit of density is T^{-2}, i.e. it does not depend on the unit of length.

Dimensions of Stress. Coefficients of Elasticity.

If f is the force exerted across an area s, the stress on s is p, where $f = ps$.

Now f is of dimensions MLT^{-2} in the C. G. S. system, and s is of dimensions L^2.

Therefore p is of dimensions $ML^{-1}T^{-2}$, since $f = ps$ is a homogeneous relation.

If t is a strain (e. g. dilatation or shear) corresponding to the stress p, k the corresponding coefficient of elasticity (e. g. resistance to compression or distortion), $p = kt$.

Since t is of no dimensions in mass, length, or time, k is of the same dimensions as a stress.

Similarly, energy (potential or kinetic) is of dimensions ML^2T^{-2}, and superficial energy being energy per unit surface is of dimensions MT^{-2}.

A couple is of dimensions ML^2T^{-2}.

§ 2. Dynamically similar systems.

Two systems of bodies are said to be dynamically similar when the numerical data which define one system can by transformation of the fundamental units be transformed into corresponding data defining the other system. Thus they are similar systems constructed on different scales.

The consideration of the conditions of dynamical similarity often leads to important results by simple methods. The following applications illustrate this :—

(1) The simple pendulum.

Let l be the length of a pendulum, g the acceleration of gravity. By altering the unit of length in the ratio $l : l'$ and the unit of acceleration in the ratio $g : g'$, the length and acceleration become l', g' respectively.

Since $T = L^{\frac{1}{2}}A^{-\frac{1}{2}}$, the unit of time is altered in the ratio

$$\sqrt{\frac{l}{l'}} : \sqrt{\frac{g}{g'}}.$$

Hence if the time of vibration was formerly denoted

by t, it is now t', where $t : t' :: \sqrt{\frac{l}{g}} : \sqrt{\frac{l'}{g'}}$.

But t' is the time of vibration of a pendulum of length l' where the acceleration of gravity is g', oscillating through the same angle as the given pendulum.

Hence the time of vibration of a pendulum of length l is proportional to $\sqrt{\frac{l}{g}}$, g being the acceleration of gravity.

The method fails to give the constant factor 2π in the complete formula, and it gives no indication that the time of vibration depends on the amplitude of oscillation.

(2) Resistance of a wire to torsion.

Let l, r be the length and radius of a wire of rigidity n, to which a total twist θ is imparted by a couple w.

Let the unit of length be decreased to $\frac{1}{x}$ th of its given value.

The length and radius of the wire become xl and xr; the rigidity becomes $x^{-1}n$, and the couple becomes $x^2 w$.

Therefore to twist a wire of length xl and radius xr through an angle θ requires a couple $x^2 w$ if the rigidity is $x^{-1} n$; it follows from Hooke's law that a couple $x^3 w$ is required if the rigidity is n.

The total twist of a length l of the wire is $\frac{\theta}{x}$. To produce a total twist θ in this length requires a couple $x^4 w$.

Hence a wire of length l, radius xr, and rigidity n requires a couple $x^4 w$ to give it a total twist θ.

Therefore the resistance to twist varies as the fourth power of the radius (Chap. VII, § 13).

(3) Kepler's Third Law.

A planet of mass m describes an orbit round a luminary of mass m'. If the unit of length is changed from L to

$\frac{1}{x} \cdot L$, and the unit of mass from M to $\frac{1}{z} \cdot M$, the astronomical unit of time changes from T to $Tx^{-\frac{3}{2}} z^{\frac{1}{2}}$, since

$$L^3 T^{-2} = M.$$

If d is originally any diameter of the orbit, xd denotes the same diameter after the change of units; and, if t originally denotes the time of describing the orbit, $tx^{\frac{3}{2}} z^{-\frac{1}{2}}$ denotes this time after the units have been changed.

Also the periodic time does not depend on the mass of the planet, for the acceleration is independent of this mass.

Hence if the planet describes an orbit whose principal diameter is d round a luminary of mass m' in time t, it will describe a similar orbit with principal diameter xd round a luminary of mass zm' in a time $x^{\frac{3}{2}} z^{-\frac{1}{2}} t$.

Hence for *similar* orbits the periodic time is proportional to $\sqrt{\frac{d^3}{m'}}$, m' being the mass of the luminary, d the principal diameter of the orbit.

Thus we have proved Kepler's Third Law for the case of similar orbits.

(4) Consider a mass of liquid at rest under its own gravitation. It will assume the form of a sphere, and if the surface is deformed in any way and then released the liquid will be set in vibration. It is required to find the relation of the period of the vibration to the size and density of the sphere.

Let t be the time of vibration, d the density of the sphere. As the force in question is that of gravitation, we use the astronomical system of units, and in this system $D = T^{-2}$ Hence if we alter the unit of length only, the density and time of vibration are denoted by the same numbers as before. Therefore in similarly deformed liquid spheres of the same material the time of vibration is the same whatever may be the radius of the vibrating sphere. In the case of spheres of different materials we must suppose the unit of density, and therefore the unit of time, to vary. If the density denoted initially by d is finally denoted by d', a time denoted initially by t is finally denoted by $t\sqrt{\frac{d}{d'}}$, or t'.

Hence for similarly deformed spheres of different densities

$$t \propto \frac{1}{\sqrt{d}},$$

t being the time of vibration, and d the density of a sphere.

If g is the acceleration of a falling body at the surface of the sphere, and a is the radius,

$$g = \frac{4}{3}\pi d a^3 \div a^2 = \frac{4}{3}\pi d a,$$

and $$t \propto \sqrt{\frac{a}{g}}.$$

The reader will find it a useful exercise to construct a system of units analogous to the astronomical system, assuming that the law of force between two masses m, m' at a distance r apart is $\frac{m\,m'}{r^n}$, and thence to compare the periodic times of similar orbits described by m about m' under this law of force. Making $n = -1$, the isochronism of simple harmonic motions may be verified.

Example.—A particle oscillates along a straight line AB under an attraction to A which is proportional to the distance from A.

Show that the oscillations are isochronous when the motion is resisted by a force proportional to the velocity.

The above method may also be applied more briefly by noticing that all physical equations are homogeneous. Thus if the time of vibration t of a pendulum is required, it is assumed that it depends only on l the length of the pendulum, and on m and mg the mass and weight of the bob. It can therefore be expressed by the formula $Al^x m^y g^z$, when A, x, y, z are numerical constants. Now this expression is of dimensions T.

Therefore
$$x + z = 0,$$
$$y = 0,$$
$$z = -\tfrac{1}{2},$$
$$x = \tfrac{1}{2},$$
and
$$t = A\sqrt{\frac{l}{g}}.$$

In order that this method may be trustworthy we must be sure that we have enumerated all the quantities on which t (or any other quantity required) depends.

Example.—To compare the times of vibration of different liquid spheres held together by their surface-tensions, other forces being neglected.

The time of vibration depends on the surface-tension Σ of dimensions MT^{-2} and the mass M of the sphere.

Hence
$$t = \Sigma^x M^y,$$
and
$$2x = -1,$$
$$x + y = 0.$$

Therefore
$$t = A\sqrt{\frac{M}{\Sigma}},$$
where A is a numerical constant.

It must be understood that the vibrations of the spheres are similar, i.e. that the spheres are similarly deformed before being left to vibrate.

A good illustration of the care needed in applying the Principle of Dynamical Similitude is afforded by an invention which was brought before a recent Commission on Mines. It was an arrangement for preventing a cage from falling to the bottom of a shaft if by any accident the rope supporting the cage broke. A spring, compressed by a weight, would on being freed from pressure release a clutch, which after release (in the model) held the guide ropes in the shaft, and so prevented the cage from descending. If by any chance the rope broke, the cage would begin to descend with nearly the acceleration of gravity, the pressure of the spring would become very small, the clutch would be released, and the cage stopped.

In applying conclusions derived from this model to the cage in an actual shaft, it is necessary to bear in mind that the moving mass m and its velocity v are very considerably greater than the corresponding quantities in the model, and that work $\frac{1}{2}mv^2$ must be spent in bringing the cage to rest. The strain on the guides is thus very considerable indeed compared with that in the model, and correct inferences could only be drawn if the guides in the model were made correspondingly weak.

ANSWERS TO EXAMPLES

CHAPTER I.

Page 10. Ex. 5. $2\sqrt{2}$ along DB. 6. 0.

8. (1) $a(\sqrt{2}-1)$ parallel to BA.
(2) $a\sqrt{3}$, making an angle $\tan^{-1}\sqrt{2}$ with AB.

Page 11. 2. 33 ft. per second.

Page 22. 4. ·00014 ... inches per second. ·000156...
5. 5 miles an hour, 330 yards.

Page 29. 3. 100. 4. -65. 5. 172, 208, 244.
6. 28, -8, -44. 7. $33\frac{1}{3}$. 8. $-55\frac{5}{9}$.
9. 200. 10. 65. 11. 625, $-137·5$. 12. 8.
13. 12·5. 14. -8, 40; 36, 28, 20, 12, 4.
15. 16, 64, 144. The distance travelled in the 10th second is 304 ft. 16. $4\frac{5}{8}$. $\frac{8}{13}$ sec.
17. 6. $\frac{5}{8}$ sec. 18. $107\frac{1}{3}$, $429\frac{1}{3}$.

Page 34. 4. 6 secs.; 176·4 metres.
5. (1) 5 seconds;
(2) 122·5 metres;
(3) 7840 centimetres per second.

7. 160. 8. 522.

9. $48\sqrt{2}$ at an angle $45°$ from the horizon, 3 secs.

12. 60·08 ... feet per second.

Page 42. 4. ·01.

5. 60 miles an hour, 612 yards. 8 seconds.

6. ·125. 7. 28800 ft. $960\sqrt{2}$.

8. ·00007295...

9. $·893... \times 10^{-9}$, $·1931... \times 10^{-6}$.

CHAPTER II.

Page 78. 9. $\frac{7^5 \times 5^4}{4}$, $\frac{7^8 \times 5^2}{2}$.

10. 545 cm. per sec., $\frac{5}{9}$ sec., $13352\frac{1}{2}$.

13. (1) 468 grams weight, 195;
(2) 432,75; (3) 180.

15. $12P$. 17. $150\sqrt{2}-100$ along AC.

18. 300 poundals. 19. $3\sqrt{3}$, 3.

20. 8, −32, 96, 24. 21. $5\frac{1}{3}$ ft. 22. 240, 960.

23. $\frac{2}{3}g$, $3\frac{1}{3}$ lbs. weight, $\frac{3}{8}$ sec. 24. 214·8...

25. 984·94. 26. 20 ft. per sec.

27–29. See § 11. 30. $\frac{g}{4\pi^2}$.

31. 16 ft., $10\frac{2}{3}$. 32. 8 ft.

33. (1) $9\frac{3}{8}$ in.; (2) $9\frac{3}{8}$ lbs.

CHAPTER III.

Page 83. 1. 2,060,100 ergs. (3) 40·24.

Page 109. 4. 22500 foot-poundals. 5625 poundals.

5. 160 lbs. weight in each rope. 6. $\frac{2}{\pi}$.

7. $96\frac{1}{4}$ min. 8. See Ex. 2.

9. $273\frac{2}{9}$ H.P. 10. $\frac{1}{10}$. $\frac{6t}{275}$.

11. At a height $120\sqrt{3}-64$ ft.

After 2 sec. 60 ft. from the wall.

12. The subsequent velocities are -16 and 14.

13. 24.1... 14. 108.

15. 3,080,000, $26\frac{2}{3}$ min. 17. $e^2 = 2 - \sqrt{2}$.

CHAPTER IV.

Page 138, 1. 6 lbs. weight.

7. (1) $\frac{29}{64}Ma^2$;

(2) $\frac{113}{1600}Ma^2$, M being the mass, and a the radius.

8. $m\sqrt{\frac{lg}{3}}$.

9. If W is the weight of the bar, the tension is $\frac{3W}{\sqrt{2}}$, and the force at the hinge is $\frac{W\sqrt{10}}{2}$.

11. $\frac{2048}{17\pi}$. 13. $\frac{9800}{3}\pi^2$ ft. poundals.

14. $56000\pi^2$, $233\frac{1}{3}\pi$.

15. $\frac{30625}{512}\pi^2$. 16. 1000.

CHAPTER V.

Page 175. 6. 7·4; 6·6.

7. $10\sqrt{73}$; $\frac{3}{8}$.

8. $\frac{10}{3\sqrt{3}}$ lbs. weight.

10. $6\,\frac{2+\sqrt{3}}{2\sqrt{3}+1}$ lbs. weight.

11. $\frac{5}{4\sqrt{2}}$ cwt.

14. $\frac{2a}{b}$.

17. $\frac{700}{9\sqrt{5}}$. $\frac{550}{9\sqrt{5}}$

18. $32\frac{1}{2}$ lbs. weight.

19. $T(\cot\alpha + \mu) = \frac{1}{2}W$, W being the weight of the beam.

CHAPTER VIII.

Page 269. 6. 14000 lbs. weight; $4\sqrt{2}$ ft. from the surface.

8. $\frac{14688\pi}{35}$ lbs.

9. To $15\sqrt{2}$ inches.

10. 16000 lbs. weight; 20000; $5\frac{1}{3}$; $5\frac{1}{15}$ ft. below the upper edge of the square.

11. $\sqrt[3]{\frac{2\rho-1}{2\rho}}$.

12. 4 : 13.

13. $21\frac{1}{3}$ lbs. per sq. inch.

14. $6\frac{11}{13}$ cm.

15. (1) 0·95; (2) 1.

16. 8·4 in. below the top, $15\frac{5}{8}$ lbs. weight.

17. 774 : 932.

19. $\frac{125}{288}$, $\frac{125\pi}{18}$ lb. weight.

20. $\frac{875\pi}{16}$, $\frac{3125\pi}{8}$.

21. 10 : 13.

22. 2·5.

23. volume of lead : volume of sulphur : : 1 : 10·4.

25. 7 : 1.

www.ingramcontent.com/pod-product-compliance
Lightning Source LLC
LaVergne TN
LVHW021939220826
846092LV00010B/1177

* 9 7 8 9 3 5 5 2 8 1 8 9 0 *